T0213785

Communications in Computer and Information Science 637

Commenced Publication in 2007
Founding and Former Series Editors:
Alfredo Cuzzocrea, Dominik Ślęzak, and Xiaokang Yang

More information about this series at http://www.springer.com/series/7899

Mirjana Ivanović · Bernhard Thalheim
Barbara Catania · Klaus-Dieter Schewe
Mārīte Kirikova · Petr Šaloun
Ajantha Dahanayake · Tania Cerquitelli
Elena Baralis · Pietro Michiardi (Eds.)

New Trends in Databases and Information Systems

ADBIS 2016 Short Papers and Workshops, BigDap, DCSA, DC
Prague, Czech Republic, August 28–31, 2016
Proceedings

 Springer

Editors

Mirjana Ivanović
Faculty of Sciences
University of Novi Sad
Novi Sad
Serbia

Bernhard Thalheim
Christian-Albrechts-Universität Kiel
Kiel
Germany

Barbara Catania
University of Genoa
Genoa
Italy

Klaus-Dieter Schewe
Software Competence Center Hagenberg
  GmbH
Hagenberg
Austria

Mārīte Kirikova
Riga Technical University
Riga
Latvia

Petr Šaloun
VSB-Technical University Ostrava
Ostrava
Czech Republic

Ajantha Dahanayake
Georgia College and State University
Milledgeville, GA
USA

Tania Cerquitelli
Politecnico di Torino
Torino
Italy

Elena Baralis
Politecnico di Torino
Torino
Italy

Pietro Michiardi
EURECOM
Biot Sophia Antipolis cedex
France

ISSN 1865-0929 ISSN 1865-0937 (electronic)
Communications in Computer and Information Science
ISBN 978-3-319-44065-1 ISBN 978-3-319-44066-8 (eBook)
DOI 10.1007/978-3-319-44066-8

Library of Congress Control Number: 2016947192

Printed on acid-free paper

This Springer imprint is published by Springer Nature
The registered company is Springer International Publishing AG Switzerland

Mirjana Ivanović · Bernhard Thalheim
Barbara Catania · Klaus-Dieter Schewe
Mārīte Kirikova · Petr Šaloun
Ajantha Dahanayake · Tania Cerquitelli
Elena Baralis · Pietro Michiardi (Eds.)

New Trends in Databases and Information Systems

ADBIS 2016 Short Papers and Workshops, BigDap, DCSA, DC
Prague, Czech Republic, August 28–31, 2016
Proceedings

 Springer

Editors

Mirjana Ivanović
Faculty of Sciences
University of Novi Sad
Novi Sad
Serbia

Bernhard Thalheim
Christian-Albrechts-Universität Kiel
Kiel
Germany

Barbara Catania
University of Genoa
Genoa
Italy

Klaus-Dieter Schewe
Software Competence Center Hagenberg
   GmbH
Hagenberg
Austria

Mārīte Kirikova
Riga Technical University
Riga
Latvia

Petr Šaloun
VSB-Technical University Ostrava
Ostrava
Czech Republic

Ajantha Dahanayake
Georgia College and State University
Milledgeville, GA
USA

Tania Cerquitelli
Politecnico di Torino
Torino
Italy

Elena Baralis
Politecnico di Torino
Torino
Italy

Pietro Michiardi
EURECOM
Biot Sophia Antipolis cedex
France

ISSN 1865-0929   ISSN 1865-0937 (electronic)
Communications in Computer and Information Science
ISBN 978-3-319-44065-1   ISBN 978-3-319-44066-8   (eBook)
DOI 10.1007/978-3-319-44066-8

Library of Congress Control Number: 2016947192

Printed on acid-free paper

This Springer imprint is published by Springer Nature
The registered company is Springer International Publishing AG Switzerland

Preface

The 20th East-European Conference on Advances in Databases and Information Systems (ADBIS 2016) took place in Prague, Czech Republic, during August 28–31, 2016. The ADBIS series of conferences aims at providing a forum for the dissemination of research accomplishments and at promoting interaction and collaboration between the database and information systems research communities from Central and East European countries and the rest of the world. The ADBIS conferences provide an international platform for the presentation of research on database theory, development of advanced DBMS technologies, and their advanced applications. As such, ADBIS has created a tradition: its 20th anniversary edition in 2016 continued the ADBIS series held in St. Petersburg (1997), Poznan (1998), Maribor (1999), Prague (2000), Vilnius (2001), Bratislava (2002), Dresden (2003), Budapest (2004), Tallinn (2005), Thessaloniki (2006), Varna (2007), Pori (2008), Riga (2009), Novi Sad (2010), Vienna (2011), Poznan (2012), Genoa (2013), Ohrid (2014), and Poitiers (2015). The conferences are initiated and supervised by an international Steering Committee consisting of representatives from Armenia, Austria, Bulgaria, Czech Republic, Cyprus, Estonia, Finland, France, Germany, Greece, Hungary, Israel, Italy, Latvia, Lithuania, FYR of Macedonia, Poland, Russia, Serbia, Slovakia, Slovenia, and the Ukraine.

This volume contains 11 papers selected as short contributions to be presented at the ADBIS conference as well as papers contributed by two associated satellite events and a Doctoral Consortium. Each of the satellite events and Doctoral Consortium complementing the main ADBIS conference had its own international Program Committee, whose members served as the reviewers of papers included in this volume. The volume is divided into four parts, one devoted to ADBIS short contributions and each other part to a single satellite event and Doctoral Consortium.

This volume also contains a survey on the ADBIS history written by Theodora Tsikrika and Yannis Manolopoulos. ADBIS has had a pre-history in the East–West symposia and symposia of the ACM SIGMOD Chapter in Moscow. The ADBIS conference series is now 20 years old and can be considered one of the most established and well-recognized conferences in the field of database and information systems technology, theory, practice, and modeling. Although in the beginning the conference was an East–West forum, it has now become a worldwide conference. ADBIS has not focused on specific sub-areas of database and information systems research.

The selected short papers span a wide spectrum of topics related to the ADBIS conference. Most of them are related to database and information systems technology for advanced applications. Typical applications are big data, multidimensional data, novel data store techniques, streaming databases, NoSQL applications, crowdsourcing support systems, data integration and evolution, user-oriented systems, and data warehouses.

The workshop co-chairs Barbara Catania and Klaus-Dieter Schewe selected three of the submitted workshop proposals for ADBIS 2016 with the condition that each of the workshops would have at least six submissions. Two workshops met this condition. One workshop got cancelled.

The Third International Workshop on Big Data Applications and Principles (BigDap 2016) was organized by Tania Cerquitelli (Politecnico di Torino, Italy), Elena Baralis (Politecnico di Torino, Italy), and Pietro Michiardi (EURECOM, France).

In the last few years, a variety of modern applications, e.g., computer network traffic, smart cities, e-commerce, social networks, are able to generate data at an unprecedented rate, to such an extent that data rapidly scale toward big data. These large data volumes provide an unprecedented opportunity to tackle interesting research challenges so as to add intelligence in real-world applications. The importance of effectively and efficiently dealing with big data collections is revealed by the growing number of companies engaged in the field.

The BigDap 2016 workshop included a keynote presentation held by Marco Mellia on the design and investigation of energy-efficient networks (green networks) and in traffic monitoring and analysis. The Internet is based on decentralization and diversity and its distributed nature leads to operational brittleness and difficulty in identifying the root causes of performance degradation. In such a context, network measurements are a fundamental pillar for shedding light and unveiling design and implementation defects. The mPlane runs, collects, and analyzes traffic measurements to study the operation and functioning of the Internet. The potentiality of the mPlane approach to unveil network and service degradation issues in live and operational networks is discussed.

The workshop included an invited paper (Garza et al.) and three selected research papers (Ordozgoiti et al., Venturini et al., Zhu and Mozo). In the invited paper, a context-aware personalization approach is proposed to efficiently address the information overload issue. Data mining algorithms have been exploited to automatically infer contextual views over a relational database.

The three selected papers address different interesting research issues related to innovative algorithms on big data management and analysis. An unsupervised feature selection algorithm, based on data orthogonal transformations together with rank revealing matrix factorizations, was proposed by Ordozgoiti et al. to efficiently perform cluster analysis on large data collections. A novel distributed associative classifier (named BAC), based on ensemble techniques, was presented by Venturini et al. BAC, designed and developed on the Apache Spark framework, building a variety of models on different subsets of the original dataset and exploiting a voting strategy to provide a unique classification outcome. Zhu and Mozo proposed the Spark2Fires algorithm to efficiently perform subspace clustering on large data sets. The corresponding Apache Spark implementation was evaluated and discussed.

The Second Workshop on Data-Centered Smart Applications (DCSA 2016) sought to bring together computer science and information systems experts, and more precisely data scientists, involved in smart applications engineering. Massive data collections and the usage of that data for smart applications has become key to the process of improving the efficiency, reliability, and security of a traditional application. There are many issues that should be taken into consideration as the first step: (a) modeling, analytics and design of data and smart applications; (b) metadata, ontologies, vocabularies perspectives; (c) Semantic Web for smart applications; (d) applications of existing technologies, etc. The classic approach where people had to learn how

equipment can be used is no longer appropriate. Applications are only accepted if they are natural to the users. In the past, researchers tried participatory design based on some knowledge of the main users, their way of operating systems, and their main desires and demands. This research and technology development must be extended by tools that can be used in an intuitive way, within many different cultures, within a variety of deployment scenarios, within a group of people deploying the techniques, within different collaboration scenarios, and within different levels of attention.

Applications should be flexible to use and provide features for intelligent extraction of essential data within a given scope. Intelligent extraction of essential data and interpreting these data in a proper way is still a challenge. The invited paper (Martinez-Gil et al.) introduces such features based on a case study in a human resource management application. Smart applications must also be adaptable to the culture of the users despite their development for completely different environments. The four accepted papers were selected from ten submissions and span topics such as modeling of smart applications and modeling for collaborations, adaptation of services to cultures, and smart systems for extreme disastrous situations. E. Alsabi et al. discuss how smart applications are often developed for mobile applications and require novel development methods, e.g., on the basis of design science. Smart applications require robustness of data management also in the case that some of the data are uncertain, incomplete, or problematic. In a case study reported by K. Honda and N. Yoshida, it is shown how such data can be enhanced through integration with high-quality data. Applications have to support collaboration among users and among systems. Model-based multi-party collaboration is the topic of the paper by M. Tropmann-Frick et al. It supports applications that use a variety of models and systems. R. Ehaidid et al. show how smart applications can also be adapted to the culture of users despite their development for completely different environments.

The ADBIS Doctoral Consortium (ADBIS DC 2016) was a forum for PhD students to present their research ideas, discuss them with the scientific community, receive feedback from senior mentors, socialize, and form cooperation networks. With this purpose, the ADBIS main conference and workshops were open to ADBIS DC 2016 participants. The DC organizers chose three papers in the final selection. Their authors are Jiri Sebek and Karel Richta from the Czech Republic, Ondrej Kassak, Michal Kompan, and Maria Bielikova from Slovakia, and Stanislav Makarov from Russia. The topics of the DC were on "Adaptive Application Structure Using the Aspect-Oriented Approach," "Short-Term User Behavior Changes Modeling," "Framework for Managing Distinct Versions of Data in Relational Databases," respectively.

We would like to express our gratitude to every individual who contributed to the success of ADBIS 2016. Firstly, we thank all authors for submitting their research papers to the conference. However, we are also indebted to the members of the community who offered their precious time and expertise in performing various roles ranging from organizational to reviewing roles – their efforts, energy, and degree of professionalism deserve the highest commendations. Special thanks go to the Program Committee members and the external reviewers for their support in evaluating the papers submitted to ADBIS 2016, ensuring the quality of the scientific program. We also thank all the colleagues, secretaries, and engineers involved in the conference and workshops organization, particularly Milena Zeithamlova (Action M Agency) for her endless help and support. A special thank you to the members of the Steering Committee, and in particular, its chair, Leonid Kalinichenko, and vice chair, Yannis Manolopoulos, for all their help and guidance.

BigDAP 2016 was sponsored by the European Union under the FP7 Grant Agreement n. 619633 ("ONTIC" Project – Online Network Traffic Characterization).

The conference would not have been possible without our supporters and sponsors: the Faculty of Mathematics and Physics (Charles University in Prague), VSB – Technical University of Ostrava, Czech Society for Cybernetics and Informatics (CSKI), and the software companies Profinit, DCIT and INTAX.

Finally, we thank Springer for publishing the proceedings containing invited and research papers in the CCIS series. The Program Committee work relied on EasyChair, and we thank its development team for creating and maintaining it; it offered great support throughout the different phases of the reviewing process.

June 2016

Mirjana Ivanović
Bernhard Thalheim
Barbara Catania
Klaus-Dieter Schewe
Mārīte Kirikova
Petr Šaloun
Ajantha Dahanayake
Tania Cerquitelli
Elena Baralis
Pietro Michiardi

Organization

Program Committee

Witold Abramowicz	Poznan University of Economics, Poland
Bader Albdaiwi	Kuwait University
Birger Andersson	Royal Institute of Technology
Grigoris Antoniou	University of Huddersfield, UK
Costin Badica	University of Craiova, Romania
Marko Bajec	University of Ljubljana, Slovenia
Ladjel Bellatreche	ISAE - ENSMA
Andras Benczur	Eotvos Lorand University, Hungary
Maria Bielikova	Slovak University of Technology in Bratislava, Slovakia
Alexander Bienemann	Christian-Albrechts-Universität zu Kiel, Germany
Miklos Biro	Software Competence Center Hagenberg, Austria
Zoran Bosnic	University of Ljubljana, Slovenia
Doulkifli Boukraa	University of Jijel, Algeria
Drazen Brdjanin	University of Banja Luka, Bosnia and Herzegovina
Stephane Bressan	National University of Singapore
Bostjan Brumen	University of Maribor, Slovenia
Zoran Budimac	University of Novi Sad
Albertas Caplinsks	Institute of Mathematics and Informatics
Barbara Catania	DIBRIS-University of Genoa, Italy
Krzysztof Cetnarowicz	AGH - University of Science and Technology of Krakow, Poland
Ajantha Dahanayake	Georgia College and State University, USA
Antje Duesterhoeft	University of Applied Sciences, Wismar, Germany
Johann Eder	Alpen Adria Universität Klagenfurt, Austria
Erki Eessaar	Tallinn University of Technology, Estonia
Markus Endres	University of Augsburg, Germany
Werner Esswein	Technical University of Dresden, Germany
Georgios Evangelidis	University of Macedonia, Thessaloniki, Greece
Flavio Ferrarotti	Software Competence Center Hagenberg (SCCH), Germany
Peter Fettke	Institute for Information Systems at DFKI (IWi), Germany
Peter Forbrig	University of Rostock, Germany
Flavius Frasincar	Erasmus University Rotterdam, The Netherlands
Dirk Frosch-Wilke	University of Applied Sciences Kiel, Germany
Jan Genci	Technical University of Kosice, Slovakia

Janis Grabis	Riga Technical University, Latvia
Gunter Grafe	HTW Dresden
Giancarlo Guizzardi	Federal University of Espírito Santo (UFES), Brazil
Hele-Mai Haav	Tallinn University of Technology, Institute of Cybernetics, Estonia
Theo Haerder	TU Kaiserslautern, Germany
Mirjana Ivanović	University of Novi Sad, Serbia
Hannu Jaakkola	Tampere University of Technology, Finland
Stefan Jablonski	University of Bayreuth, Germany
Klaus P. Jantke	Fraunhofer IDMT, Germany
Leonid Kalinichenko	Institute of Informatics Problems RAS, Russia
Ahto Kalja	Tallinn University of Technology, Estonia
Mehmed Kantardzic	University of Louisville, USA
Dimitris Karagiannis	University of Vienna, Austria
Zoubida Kedad	University of Versailles, France
Mikhail Kogalovsky	Market Economy Institute of the Russian Academy of Sciences, Russia
Michal Kopecky	Charles University in Prague
Michal Kratky	VSB-Technical University of Ostrava, Czech Republic
John Krogstie	IDI, NTNU
Wolfgang Lehner	Technical University Dresden, Germany
Sebastian Link	University of Auckland, New Zealand
Audrone Lupeikiene	Vilnius University, Lithuania
Hui Ma	Victoria University of Wellington, New Zealand
Leszek Maciaszek	Wrocław University of Economics
Christian Mancas	Ovidius University, Romania
Yannis Manolopoulos	Aristotle University of Thessaloniki, Greece
Rainer Manthey	University of Bonn, Germany
Manuk Manukyan	Yerevan State University, Armenia
Karol Matiasko	University of Zilina, Slovakia
Brahim Medjahed	University of Michigan - Dearborn, USA
Dezso Miklos	Hungarian Academy of Sciences, Hungary
Pavle Mogin	Victoria University of Wellington, New Zealand
Tadeusz Morzy	Poznan University of Technology, Poland
Pavol Navrat	Slovak University of Technology, Slovakia
Martin Necasky	Charles University in Prague
Boris Novikov	St.-Petersburg University, Russia
Andreas Oberweis	Universität Karlsruhe, Germany
Zoran Obradović	Temple University, USA
Andreas L. Opdahl	University of Bergen, Norway
George Angelos Papadopoulos	University of Cyprus
Tomas Pitner	Masaryk University, Czech Republic
Jan Platos	VSB - Technical University of Ostrava, Czech Republic
Vedran Podobnik	University of Zagreb, Croatia

Additional Reviewers

Aboelfotoh, Hosam
Baryannis, George
Batsakis, Sotiris
Berkani, Nabila
Bork, Dominik
Braun, Richard
Broneske, David
Chen, Xiao
Dosis, Aristotelis
Egert, Philipp
Emrich, Andreas

Fekete, David
Gonzalez, Senen
Hussain, Zaid
Lacko, Peter
Lechtenborger, Jens
Lukasik, Ewa
Marenkov, Jevgeni
Mehdijev, Nijat
Meister, Andreas
Mettouris, Christos
Niepel, Ludovit

Normantas, Kestutis
Peska, Ladislav
Rehse, Jana
Robal, Tarmo
Rossler, Richard
Schomm, Fabian
Stupnikov, Sergey
Tec, Loredana
Zierenberg, Marcel

Contents

BigDap 2016 – Big Data Applications and Principles

DCSA 2016 – Data Centered Smart Applications

ADBIS Doctoral Consortium

A Retrospective Study on the 20 Years of the ADBIS Conference

Theodora Tsikrika[1] and Yannis Manolopoulos[2(✉)]

[1] Centre for Research and Technology Hellas, 57001 Thessaloniki, Greece
theodora.tsikrika@iti.gr
[2] Department of Informatics, Aristotle University, 54124 Thessaloniki, Greece
manolopo@csd.auth.gr

Abstract. The East-European Conference in Advances in Data Bases & Information Systems (ADBIS) spans 20 years of life. Here, by using simple statistic measures and bibliographic analysis, we illustrate basic characteristics and features of ADBIS, i.e. the venues, persons and countries involved. Also, its international character, its competitiveness and its status in the community are revealed. Finally, prolific authors and countries are extracted, topics of interest are deduced, collaboration issues and citing performance are illuminated.

1 Introduction

In September 1992, the Moscow ACM SIGMOD Chapter was chartered, chaired by Leonid Kalinichenko. Immediately after its formation, the Chapter founded the Annual Workshops "Advances in Data Bases & Information Systems" (ADBIS). In May 1993, the ADBIS workshop was organised in Moscow under the subtitle "1st Joint Workshop of the Kiev and Moscow ACM SIGMOD Chapters". During 1994–1996, three ADBIS Workshops were organised in Moscow by the Moscow ACM SIGMOD Chapter under the subtitle "International Workshop". In 1996, after discussions with the ACM SIGMOD (mostly with its then Chair, Won Kim), it was decided to transform ADBIS into an East European forum for the exchange of scientific advancements in databases and information systems between the research communities in Eastern Europe and the rest of the world. Thus, for the next two years ADBIS was organised as an "East European Symposium", whereas since 1999 its status is as an "East European Conference".

This year the ADBIS conference becomes 20 years old since it was established as an East European event in 1997. In the meantime, the ADBIS conference has attracted the international interest of the research community and is being mentioned in several ranking lists and indexed in several digital libraries, such as DBLP[1], Google Scholar, Microsoft Academic Search and so on. In this paper, by using simple measures and bibliographic analysis, we illustrate basic characteristics and features of ADBIS, i.e. the venues, persons and countries involved.

---

[1] Database Systems and Logic Programming, http://dblp.uni-trier.de/db/conf/adbis/.

© Springer International Publishing Switzerland 2016
M. Ivanović et al. (Eds.): ADBIS 2016, CCIS 637, pp. 1–15, 2016.
DOI: 10.1007/978-3-319-44066-8_1

Also, its international character, competitiveness and status in the community are revealed. Finally, prolific authors and countries are extracted, topics of interest are deduced, collaboration issues and citing performance are illuminated.

2   Key-Persons and Venues

The ADBIS Conference is managed by a Steering Committee, which initially (1997) was comprised by 13 persons from 11 countries and expanded until 2016 to 28 persons from 19 countries (see Table 1). In addition, Table 2 depicts for each year the venue, the main key organisers along with the invited persons and their countries. We notice that General Chairs and PC chairs are affiliated with 21 countries, whereas the invited speakers are affiliated with 21 countries as well. These figures indicate the sturdy international foundations of ADBIS. In the sequel, additional information will be provided to support this argument.

Table 1. Members of the ADBIS Steering Committee

Steering Committee 1997		
R. Bercaru (RO)	L. Kalinichenko (RU) *Chair*	B. Novikov (RU)
A. Čaplinskas (LT)	M. Kogalovsky (RU)	A. Stogny (UA)
J. Eiduks (LV)	T. Morzy (PL)	T. Weltzer (SI)
H.-M. Haav (EE)	P. Navrát (SK)	V. Wolfengagen (RU)
	J. Pokorný (CZ)	
Steering Committee 2016		
P. Atzeni (IT)	H. Jaakkola (FI)	M. Nikitchenko (UA)
A. Benczúr (HU)	L. Kalinichenko (RU) *Chair*	J. Pokorný (CZ)
A. Čaplinskas (LT)	M. Kogalovsky (RU)	B. Rachev (BG)
B. Catania (IT)	Y. Manolopoulos (GR) *Vice-chair*	B. Thalheim (DE)
J. Eder (AT)	R. Manthey (DE)	G. Vossen (DE)
T. Härder (DE)	M. Manukyan (AM)	T. Weltzer (SI)
M. Kirikova (LV)	J. Michaeli (IL)	V. Wolfengagen (RU)
H.-M. Haav (EE)	T. Morzy (PL)	R. Wrembel (PL)
M. M. Ivanović (RS)	P. Navrát (SK)	E. Zumpano (IT)
	B. Novikov (RU)	

In the early years (until 2002), ADBIS events were organised in cooperation with ACM SIGMOD. ADBIS joined the ACM SIGMOD program "SIGMOD Digital Symposium Collection (DiSC)" aiming at collecting symposium anthologies on CDs. More information about this cooperation can be found in [2].

Since 1998 the ADBIS proceedings are published by Springer in the Lecture Notes in Computer Science (LNCS) series. In addition, a second volume with

Table 2. Venue and key persons during the 20 years of ADBIS

Year-Venue	General Chairs	PC Chairs	Invited speakers
1997-St. Petersburg		R. Manthey (DE) W. Wolfengagen (RU)	H. Garcia-Molina (USA)
1998-Poznan	T. Morzy (PL)	W. Litwin (FR) G. Vossen (DE)	T. Imieliński (USA) C. Mohan (USA)
1999-Maribor	I. Rozman (SI)	J. Eder (AT) T. Welzer (SI)	S. Alagic (USA) E. Neuhold (DE) G. Pernul (DE)
2000-Prague	J. Pokorný (CZ)	Y. Masunaga (JP) J. Štuller (CZ) B. Thalheim (DE)	S. Nishio (JP) S. Spaccapietra (CH) H. Schweppe (DE)
2001-Vilnius	A. Zavadskas (LT)	J. Eder (AT) A.Čaplinskas (LT)	F. Garzotto (IT) L. Kalinichenko (RU) J. Schmidt (DE)
2002-Bratislava	L. Molnár (SK)	Y. Manolopoulos (GR) P. Navrát (SK)	P. Atzeni (IT) O. Guenther (DE) H.-J. Schek (CH)
2003-Dresden	B. Thalheim (DE) U. Wloka (DE)	L. Kalinichenko (RU) R. Manthey (DE)	F. Bry (DE) G. Bussler (DE)
2004-Budapest	A. Benczúr (HU)	J. Demetrovics (HU) G. Gottlob (AT)	J. Gray (USA) P. Revesz (USA)
2005-Talinn	J. Penjam (EE) A. Kajla (EE)	J. Eder (AT) H.-M. Haav (EE)	T. Härder (DE) N. Guarino (IT)
2006-Thessaloniki	Y. Manolopoulos (GR)	J. Pokorný (CZ) T. Sellis (GR)	S. Abiteboul (FR) Y. Ioannidis (GR) P. Zezula (CZ)
2007-Varna	B. Rachev (BG)	Y. Ioannidis (GR) B. Novikov (RU)	P. Atzeni (IT) T. Sellis (GR) G. Weikum (DE)
2008-Pori	H. Jaakkola (FI)	P. Atzeni (IT) A. Čaplinskas (LT)	H. Mannila (FI) Y. Mattias (IL) T.Özsu (USA)
2009-Riga	J. Grundspeŋķis (LV)	T. Morzy (PL) G. Vossen (DE)	M. Brantner (DE)
2010-Novi Sad	M. Ivanović (RS)	B. Catania (IT) B. Thalheim (DE)	G. Antoniou (GR) W. Cellary (PL) S. Krčo (RS) S. Rizzi (IT)
2011-Vienna	A. Min Tjoa (AT)	J. Eder (AT) M. Bielikova (SK)	M. Dumas (EE) G. Gottlob (UK) M. Henzinger (AT)
2012-Poznan	T. Morzy (PL)	T. Härder (DE) R. Wrembel (PL)	Y. Ioannidis (GR) M. Middelfart (USA)
2013-Genoa	B. Catania (IT)	J. Pokorný (CZ) G. Guerrini (IT)	A. Ailamamki (CH) M. Theobald (DE) N. Stantic (AU)
2014-Ohrid	M. Kon-Popovska (FYROM)	Y. Manolopoulos (GR) G. Trajevski (USA)	J. Gamma (PT) M. Garofalakis (GR) M. de Rijke (NL)
2015-Poitiers	L. Bellatreche (FR)	P. Valduriez (FR) T. Morzy (PL)	S. Abiteboul (FR) J. Dittrich (DE)
2016-Prague	J. Pokorný (CZ)	M. Ivanović (RS) B. Thalheim (DE)	A. Gal (IL) E. Rahm (DE) P. Zezula (CZ)

additional papers is published either by a local academic publisher, or by CEUR[2], or lastly in the AISC (Advances in Intelligent Systems & Computing) and CCIS (Communications in Computer & Information Science) series of Springer. Next, we focus on the main ADBIS proceedings and their bibliometric analysis.

3   Bibliometric Analysis Method

Bibliometric studies provide a quantitative and qualitative indication of the (scholarly) impact of research activities by analysing their associated publications and citations. In the field of Database & Information Systems, past investigations have analysed the publication behaviour [3] and citation frequencies [9] in major venues (journals and conferences) in the Database community and have reported on the impact of benchmarking activities in Information Retrieval, such as TRECVid[3] [10], CLEF[4] [11] and ImageCLEF[5] [12]. For ADBIS, a study with statistical findings about its publications and citations was conducted for the period 1994–2006 [7] to mark the first 10 years under its current status.

Bibliometric studies typically follow three steps: (i) publication data collection, (ii) citation data collection, and (iii) data analysis. Regarding the publication data, the complete lists of ADBIS proceedings can be obtained from bibliographic data sources, such as DBLP. To this end, the DBLP XML dump [6] created on May 6, 2016 was downloaded[6] and processed using XQuery queries to extract the DBLP records of the publications in the ADBIS proceedings. As author names may appear under several variations (e.g. Apostolos Papadopoulos vs. Apostolos N. Papadopoulos), the DBLP person records [6] were employed to automatically map all name variants to the "primary name" of each author, as this is identified by DBLP. Then, authors' names were manually examined to address all cases not currently identified by DBLP. In particular, authors with the same last name were examined to ensure that no duplicates existed due to variations in their first name (e.g. V. Kumar vs. Vijay Kumar).

Regarding the citation data, the most comprehensive citation data sources are: (i) Web of Science[7], (ii) Scopus[8], and (iii) Google Scholar[9]. Each follows a different data collection policy that affects both the publications covered and the number of citations found. Differences in their coverage may affect the assessment of scholarly impact metrics; the degree to which this happens varies among disciplines [1,5]. For computer science, where publications in peer-reviewed conference proceedings are highly valued and cited in their own right, the Web of Science greatly underestimates the number of citations found [1,9], given that

---

[2] Sun SITE Central Europe, http://ceur-ws.org/.
[3] TREC Video Retrieval Evaluation, http://trecvid.nist.gov/.
[4] Cross-Language Evaluation Forum, http://www.clef-initiative.eu/.
[5] CLEF Image Retrieval Evaluation, http://www.imageclef.org/.
[6] http://dblp.uni-trier.de/xml.
[7] http://apps.webofknowledge.com/.
[8] http://www.scopus.com/.
[9] http://scholar.google.com/.

its coverage of conference proceedings is very partial. Scopus and Google Scholar offer broader coverage and were both employed in this work.

The citations were obtained as follows in a 24-hour period in May 2016. In Scopus, the query "CONFNAME (adbis) AND SRCTITLE (lecture notes in computer science)" was entered in the Advanced Search and the results were cross-checked against the DBLP publication lists. The publications that could not be retrieved in response to the aforementioned query were obtained by using the query "SRCTITLE (lecture notes in computer science) AND VOLUME (*volume_number*)" again in the Advanced Search, where *volume_number* is the volume of the corresponding LNCS proceedings. The Google Scholar citation data collection was performed through *Publish or Perish (PoP)*[10]. In PoP, the "Advances in Databases and Information Systems" and "ADBIS" queries were both used in the Publication field, while the Year of publication was set between 1997 and 2016. In addition, the query "Current Issues in Databases and Information Systems" was also used in the Publication field for the particular Year 2000. The results were manually refined by removing duplicate entries and false positive matches and by merging entries deemed equivalent.

An initial examination revealed that Scopus does not cover the 1997 ADBIS proceedings, probably because these were not published by LNCS. Moreover, the number of citations varies greatly between Scopus and Google Scholar, with the latter finding around ten times more citations than Scopus (similarly to our earlier studies [11,12]). As a result, this study employs Google Scholar (and in particular its PoP wrapper) as a citation data source. Scopus is employed in a different capacity, as a supplementary publication data source. More specifically, it is used for obtaining the affiliation data of the authors of the ADBIS publications since such affiliation data are very incomplete in DBLP (in our case they are available for around 20 % of the ADBIS authors), whereas Scopus offers complete listings since it obtains them directly from the LNCS proceedings.

Finally, the analysis was performed using appropriate XQuery queries and R scripts along several axes, such as publications and citations, so as to identify trends over time and provide insights into the social network constructed by co-authorship relations and into the topics examined in ADBIS publications.

4   Results of the Bibliometric Analysis

4.1   ADBIS Publications

Table 3 focusses on the main proceedings of ADBIS 1997–2016 and provides several interesting measures. The number of submitted papers fluctuates over the years, ranging from 66 in 2008 (Pori) to a record 165 in 2010 (Novi Sad). An average of 30 papers per year has been accepted in the main ADBIS proceedings over these 20 years, with the main outliers being the 57 and 50 papers accepted in 1997 (St Petersburg) and 2010 (Novi Sad), respectively, while the absolute minimum of 21 accepted papers is observed both in the 2008 (Pori) and 2016 (Prague) ADBIS conferences.

---

[10] http://www.harzing.com/pop.htm.

Table 3. ADBIS main proceedings (1997–2016)

Year-Venue	Submitted papers	Submitting countries	Accepted papers	Accepted countries	Paper % acceptance	Country % acceptance
1997-St. Petersburg			57	25		
1998-Poznan	90		25	18	27.8	
1999-Maribor	94	33	25	17	26.6	51.5
2000-Prague	115	22	32	17	27.8	77.3
2001-Vilnius	82	30	25	20	30.5	66.7
2002-Bratislava	115	35	29	18	25.2	51.4
2003-Dresden	86		29	20	33.7	
2004-Budapest	130		27	18	20.8	
2005-Tallinn	144	40	27	17	18.8	42.5
2006-Thessaloniki	126	36	29	17	23.0	47.2
2007-Varna	77	29	23	14	29.9	48.3
2008-Pori	66	21	21	17	31.8	81.0
2009-Riga	93	28	25	15	26.9	53.6
2010-Novi Sad	165	38	50	26	30.3	68.4
2011-Vienna	105	31	30	16	28.6	51.6
2012-Poznan	122	31	32	20	26.2	64.5
2013-Genoa	92	43	26	12	28.3	27.9
2014-Ohrid	82	33	26	18	31.7	54.5
2015-Poitiers	135	39	31	23	23.0	59.0
2016-Prague	84	35	21	24	25.0	68.6
Average	105	33	30	19	27.1	57.1

The acceptance rate, i.e. the number of accepted papers over the number of submitted papers, ranges from 18.8 % in 2005, (Tallinn) where the second highest number of submissions (144) was observed, to 33.7 % in 2003 (Dresden). As the acceptance rate is considered to be correlated with the quality of the accepted material and thus it is often used as an indicator of the prestige of a conference, the 27.1 % average acceptance rate indicates that ADBIS is competitive.

Again, we remark that there is steady international community paying attention to ADBIS; each year the submitted papers originate from 33 countries on average, whereas the accepted papers originate from 18 countries. In the remainder of this study, we focus our analysis on the 1997–2015 proceedings, since the 2016 proceedings were unavailable at the time of writing.

4.2   Authors in the ADBIS Proceedings

The 578[11] papers in the 1997–2015 ADBIS main proceedings were collaboratively produced by 1,065 unique authors. Figure 1 shows the cumulative number of authors, indicating a steady increase in the number of authors joining the ADBIS community each year. Figure 2 further provides the number of authors per year and their distribution across the two categories: the "newcomers" which appear for the first time in the ADBIS proceedings (upper dark part of the bar) and the "recurring" ones which have appeared before (lower light part of the bar).

---

[11] In addition to the 569 papers listed in Table 3 for 1997–2015, we also consider 9 more papers listed in DBLP for 1998: 6 short papers and 3 in the industrial track.

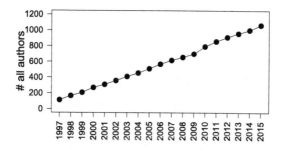

Fig. 1. Cumulative number of authors up to the indicated year

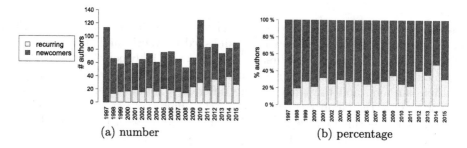

(a) number   (b) percentage

Fig. 2. Authors per year

On average, 53 new authors contribute to the accepted papers each year; the highest and lowest numbers of new authors publishing in the ADBIS proceedings (94 and 38, respectively) correspond to the conferences with the highest and lowest number of accepted papers, respectively, i.e. in 2010 (Novi Sad)[12] and 2008 (Pori). On average, 71 % of authors are newcomers despite that this percentage has dropped lastly as the ADBIS community grows. However, in any case, at least half the authors have not previously published in ADBIS.

Overall, this indicates that ADBIS is an open community that is continuously being infused with new people and thus new ideas and perspectives. This is further corroborated by Figs. 3 and 4 that depict the distributions on log-log scales of the number of papers per author and the number of years an author has published in ADBIS. Both follow a power law with the overwhelming majority of authors publishing only once in ADBIS and only about 20 % of the authors publishing multiple times. The two distributions are actually quite similar, further indicating that most authors typically publish once within each year. As a matter of fact, the maximum number of papers published within a single year by one author is three (G. Saake in 1998 and Y. Manolopoulos in 2003); 61 authors have published twice in a year, while the majority publishes only once each year.

These results further indicate that ADBIS is an open community that constantly attracts newcomers, while it also maintains its appeal to several people

---

[12] The highest number of accepted papers was actually observed in 1997. Since this was the first year of ADBIS, it is excluded, since all authors are considered as newcomers.

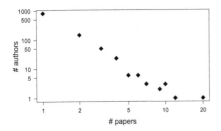

Fig. 3. Distribution of the number of papers per author

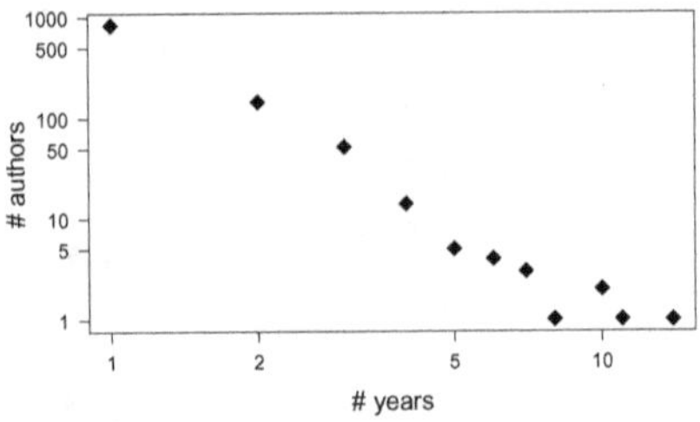

Fig. 4. Distribution of the number of years an author has published

that keep publishing in ADBIS over a number of years. Table 4 lists the top-10 authors in terms of the years they have published in ADBIS and of the number of their papers. As expected the first two columns are highly correlated, with the top-5 authors being the same in both rankings. As publications typically involve multiple authors, and thus may require different amounts of effort, inversely proportional to the number of contributors, the number of papers attributed to each author is normalised by the number of authors in each paper to achieve a fair comparison among publications with varying numbers of authors. This results in a different ranking of the most prolific authors (3rd column) that has six authors in common with the ranking based on raw counts (2nd column). Interestingly, the top-3 authors are the same in both cases. This motivates us to further examine the relations among authors in the ADBIS proceedings.

Table 4. Most engaged and most prolific authors

# years		# papers		# papers (normalised)	
Y. Manolopoulos	14	Y. Manolopoulos	20	Y. Manolopoulos	6.73
T. Härder	11	T. Härder	12	T. Härder	5.17
L.A. Kalinichenko	10	L.A. Kalinichenko	10	L.A. Kalinichenko	4.92
M. Vassilakopoulos	10	M. Vassilakopoulos	10	K. Nørvåg	4.67
W. Lehner	8	W. Lehner	10	A.V. Zamulin	3.50
A. Corral	7	T. Morzy	9	T. Morzy	3.25
T. Morzy	7	G. Saake	9	M. Vassilakopoulos	3.25
K. Subieta	7	A. Behrend	7	M. Wojciechowski	3.17
A. Behrend, J. Pokorný	6	A. Corral	7	J. Pokorný	3.08
G. Saake, M. Zakrzewicz	6	K. Subieta	7	W. Lehner	3.07

4.3   Co-authors and Their Network

An indication that there is a variation in the number of co-authors over the years has already been provided in Fig. 2a, where the highest number of authors in a year (125 in 2010 (Novi Sad)) does not correspond to the year with most accepted papers (57 in 1997 (St Petersburg)). Figure 5 shows the distribution of the number of authors per paper for each year, with the width of each "violin plot" reflecting the number of papers for a given number of authors. The line plot indicates the average number of authors which fluctuates between 2.1 and 3.3 with a slightly upward trend in recent years. This indicates a culture of collaboration, with only though a few papers (5.7 %) having five authors or more.

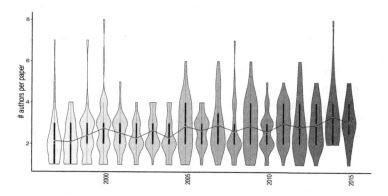

Fig. 5. Distribution of the number of authors per paper per year

Next, the co-authorship network is constructed to examine the ADBIS author community as a whole. It consists of the 1,065 authors as vertices, with 54 being isolates (i.e. authors who have not collaborated with anyone else) and 1,011 connected with 1,457 edges. The edge density is low (0.002) indicating a loosely connected community. This is further evidenced by the distribution of the size of the 226 components in the network. The largest connected component contains 78 vertices (7.7 % of all vertices) and 158 edges, while only 16 components (7.1 %) contain 10 vertices or more. Actually, more than half of the components (140 out of 226) contain only two or three vertices, likely to correspond to co-authors of papers who have not collaborated with other members of the ADBIS community.

Next, the key members of the ADBIS co-authorship network are identified based on the following centrality measures. Given an undirected network $G(N, L)$ with N nodes and L links, the *degree* of a node is the number of its neighbours, i.e. the number of co-authors in our case. The *betweenness centrality* [4] of node n_k is based on the number of paths $g_{ij}(n_k)$ from node n_i to node n_j that pass through node n_k, to the number of all paths g_{ij} from node n_i to node n_j, summed over all pairs of nodes and normalised by its maximum value $(N^2 - 3N + 2)/2$:

$$BC_k = \frac{2\sum_{i<j}^{N} \frac{g_{ij}(n_k)}{g_{ij}}}{N^2 - 3N + 2} \tag{1}$$

Betweenness centrality is, in essence, based on a broker position connecting others. Finally, Google's PageRank (PR) [8], introduced to measure the importance of Web pages, is defined for node n_k as:

$$PR_k = \frac{1-d}{N} + d \sum_{n_i \in \mathcal{N}(n_k)} \frac{PR_i}{L(n_i)} \qquad (2)$$

where d is the damping factor (set to 0.85), $L(n_i)$ is the number of links to node n_i and $N(n_k)$ is the set of nodes connected to node n_k, i.e. n_k's neighbourhood.

Table 5 lists the key members of the co-authorship network based on the three centrality measures. Y. Manolopoulos has the largest number of co-authors (26) in the ADBIS community (W. Lehner follows with 18 co-authors). This number of co-authors resulted in forming a network that corresponds to the largest connected component in the ADBIS community. Also, he has the highest betweenness centrality and PageRank values indicating his mediator role. In addition, the number of co-authors appears to be correlated with the PageRank measure, since seven authors are present in both rankings. On the other hand, the correlation appears to be weaker with the betweenness measure, as only five authors appear in both rankings. Next, the (co-)authorship of the ADBIS papers is examined in terms of the countries where the authors are based.

Table 5. Centrality measures for identifying key members of the co-authorship network

Degree		Betweeness		PageRank	
Y. Manolopoulos	26	Y. Manolopoulos	2223	Y. Manolopoulos	0.00553
W. Lehner	18	T. Morzy	1123	T. Härder	0.00427
G. Saake	15	T. Welzer	670	T. Morzy	0.00379
T. Morzy	14	A. Corral	497	W. Lehner	0.00378
C. Traina Jr	13	Y. Theodoridis	496	G. Saake	0.00336
T. Härder	12	J. Eder	297	K. Subieta	0.00336
A. J. M. Traina	11	A. Vakali	296	P. Návrat	0.00281
A. Behrend, A. Corral,	10	D. Pfoser	292	E. Pitoura	0.00267
E. Pitoura, K. Subieta,	10	B. Brumen	258	J. Pokorný	0.00247
M. Vassilakopoulos, T. Welzer	10	W. Lehner	255	L.A. Kalinichenko	0.00246

4.4 Countries

The 578 papers in the 1997–2015 ADBIS main proceedings were collaboratively produced by 1,065 unique authors affiliated with organisations in 60 countries. Figure 6 shows the cumulative number of countries, indicating an increase in the number of countries each year. As expected, this increase was more rapid in the early years and has slowed down in recent years; it has not reached a plateau yet, but it is very likely that it will increase only marginally in the future.

Table 6 lists the countries with the most solid presence over the years and also in terms of the number of papers with at least one author originating from them. Although ADBIS is supposed to be based in East Europe, it attracts

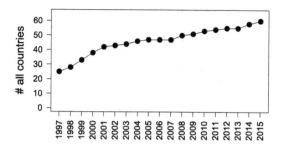

Fig. 6. Cumulative number of countries up to the indicated year

Table 6. Country statistics

# years		# papers		# papers (normalised)	
Germany	19	Germany	122	Germany	111.8
Greece	18	France	57	France	47.9
Poland	18	Greece	50	Greece	42.7
USA	18	Poland	44	Poland	40.8
France	16	USA	42	Russia	34.3
Australia	15	Russia	36	USA	30.9
Russia	15	Italy	28	Italy	25.0
Spain	14	Australia	24	Australia	18.6
Italy	13	Spain	22	Spain	17.7
United Kingdom	11	United Kingdom	15	South Korea	13.0

researchers not only from neighbouring countries (i.e. in the rest of Europe), but also from other continents as indicated by the presence of USA and Australia in the top ranks. Similarly to before, the number of papers associated with a country is normalised by taking into account the number of authors originating from that country; e.g. a paper with three authors, two from Greece and one from Spain, counts as one both for Greece and Spain in the case of raw counts (2nd column), but counts as 0.67 and 0.33, respectively, in the normalised case (3rd column). Therefore, the largest the difference between the two columns for the same country, the largest the number of its international collaborations. Overall, the rankings are the same over the top countries, with the exception of South Korea appearing instead of the United Kingdom.

Figure 7 shows the collaborations between the different countries in the ADBIS community up to 2015. It is encouraging to observe such a significant number of collaborating countries and collaborations. Moreover, the most prolific countries in terms of papers authored (i.e. Germany, France and Greece) are also (together with the USA) the most extrovert ones with the most collaborations.

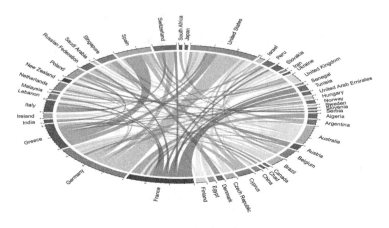

Fig. 7. Collaboration between authors in different countries in 2015

4.5 Topics

Based on the titles of the 578 papers we have extracted tag clouds with uni-grams (see Fig. 8a) and bi-grams (Fig. 8b). It is interesting to notice that the keywords depicted span a large range of database topics. Further research is necessary to evaluate the temporal evolution in this respect.

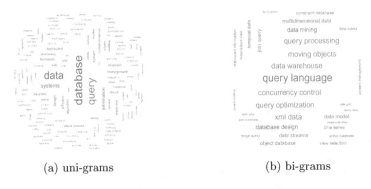

(a) uni-grams   (b) bi-grams

Fig. 8. Tag cloud extracted from the titles of the ADBIS papers

4.6 Citations

The 578 papers in the 1997–2015 ADBIS main proceedings have accumulated a total of 4,933 citations, resulting in 8.53 cites per paper on the average. As expected, the citations distribution (Fig. 9) is exponential with a long tail, with about 12 % of the papers attracting 20 citations or more and 32 % of all papers attracting 80 % of all citations. The skewness of the citation distribution is also observed by measuring its Gini coefficient, a measure of statistical dispersion that reflects the inequality among values of a frequency distribution. It corresponds to a nonnegative real number ∈ [0, 1], with higher values indicating more diverse distributions. Its overall value of 0.63 in ADBIS indicates the high degree of variability in the citations of individual publications.

Figure 10 shows the citations distribution for the papers accepted in the main proceedings each year, with the years 2000–2004 attracting the most citations in total and also including some of the most-cited papers (see also Table 7). Among the most cited, there are also papers from 1997 and 2006; as expected, more recent papers have not attracted yet significant numbers of citations. It is interesting to note that the ADBIS invited papers also have significant citations, with the 1997 Garcia-Molina et al. paper having the highest number of citations among all papers published in the ADBIS proceedings (accepted & invited).

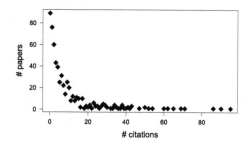

Fig. 9. Distribution of the number of citations per paper

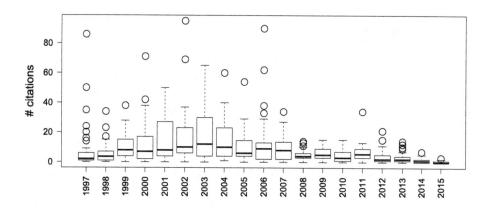

Fig. 10. Distribution of the number of citations over the years

Table 7. Most cited papers in the ADBIS proceedings

Cites	Authors	Title	Year
	Papers accepted in the main proceedings		
95	Brakatsoulas et al.	Revisiting R-tree construction principles	2002
90	Aouiche et al.	Clustering-based materialized view selection in data warehouses	2006
86	Pozewaunig et al.	ePERT: Extending PERT for workflow management systems	1997
71	Nanopoulos & Manolopoulos	Finding generalized path patterns for web log data mining	2000
69	Akal et al.	OLAP query evaluation in a database cluster: a performance study on intra-query parallelism	2002
	Invited papers in the main proceedings		
162	Garcia-Molina et al.	Semistructured Data: The TSIMMIS Experience	1997
42	Schek et al.	Infrastructure for information spaces	2002
33	Boyens & Günther	Trust is not enough: Privacy and security in ASP and Web service environments	2002

5   Conclusions

This paper has narrated how the ADBIS conference was conceived and founded in the early 90s. ADBIS has matured under its present status ("East European") as it is now 20 years old. By using statistic measures and bibliographic analysis, we illustrate basic characteristics and features of ADBIS, i.e. venues, persons and countries involved. Also, its international character, competitiveness and status in the community are revealed. Finally, prolific authors and countries are extracted, topics of interest are deduced, collaboration issues and citations are illuminated. The ADBIS family should further work towards the next 20 years.

Acknowledgements. Thanks are due to L. Kalinichenko for information on the early years of ADBIS.

References

1. Bar-Ilan, J.: Which h-index? A comparison of WoS, Scopus and Google Scholar. Scientometrics **74**(2), 257–271 (2008)
2. Ceri, S., Kalinichenko, L., Kitsuregawa, M., Lu, H., Ozsoyoglu, Z.M., Snodgras, R., Vianu, V.: SIGMOD sister societies. ACM SIGMOD Rec. **29**(1), 4–15 (2000)
3. Elmacioglu, E., Lee, D.: On six degrees of separation in DBLP-DB and more. ACM SIGMOD Rec. **34**(2), 33–40 (2005)
4. Freeman, L.C.: Centrality in social networks conceptual clarification. Soc. Netw. **1**(3), 215–239 (1978)
5. Harzing, A.-W.: Citation analysis across disciplines: The impact of different data sources and citation metrics (2010). http://www.harzing.com/data_metrics_comparison.htm. Accessed May 2016
6. Ley, M.: DBLP: some lessons learned. Proc. VLDB Endowment **2**(2), 1493–1500 (2009)
7. Manolopoulos, Y.: A statistic study for the ADBIS period 1994–2006. In: Local Proceedings of the 10th East-European Conference on Advances in Databases & Information Systems (ADBIS), pp. 247–266 (2006)
8. Page, L., Brin, S., Motwani, R., Winograd, T.: The PageRank citation ranking: bringing order to the web (1999)
9. Rahm, E., Thor, A.: Citation analysis of database publications. ACM SIGMOD Rec. **34**, 48–53 (2005)
10. Thornley, C.V., Johnson, A.C., Smeaton, A.F., Lee, H.: The scholarly impact of TRECVid (2003–2009). JASIST **62**(4), 613–627 (2011)
11. Tsikrika, T., Larsen, B., Müller, H., Endrullis, S., Rahm, E.: The scholarly impact of CLEF (2000–2009). In: Forner, P., Müller, H., Paredes, R., Rosso, P., Stein, B. (eds.) CLEF 2013. LNCS, vol. 8138, pp. 1–12. Springer, Heidelberg (2013)
12. Tsikrika, T., de Herrera, A.G.S., Müller, H.: Assessing the scholarly impact of ImageCLEF. In: Forner, P., Gonzalo, J., Kekäläinen, J., Lalmas, M., de Rijke, M. (eds.) CLEF 2011. LNCS, vol. 6941, pp. 95–106. Springer, Heidelberg (2011)

ADBIS 2016 – Short Papers

Towards Automated Performance Optimization of BPMN Business Processes

Anastasios Gounaris[(✉)]

Department of Informatics, Aristotle University of Thessaloniki, Thessaloniki, Greece
gounaria@csd.auth.gr

Abstract. Business Process Model and Notation (BPMN) provides a standard for the design of business processes. It focuses on bridging the gap between the analysis and the technical perspectives, and aims to deliver process automation. The aim of this work is to complement this effort by transferring knowledge from the related field of data-centric workflows aiming to provide automated performance optimization of the business process execution. As a key step towards this goal, the contribution of this work is to provide a methodology to map BPMNv2.0 models to annotated directed acyclic graphs, which emphasize the volume of the tokens exchanged and are amenable to existing automated optimization algorithms. In addition, concrete examples of mappings are given, while the optimization opportunities that are opened are explained.

1 Introduction

BPMN has become an international standard for designing workflows. In principle, the basic promise of BPMN is the same diagram prepared by a business analyst to be used for automating the execution of that process on a modern process engine. This however remains a vision that rarely happens in practice and the gap between the business and the technical perspectives remains [8]. As a result, the executable workflow of business processes is either manually designed in order to provide enterprize-specific configurations or derived by simple procedures using toolkits from established vendors (e.g., Bigazi, IBM, Oracle). Either way, any performance optimization responsibilities rest with experienced IT technicians.

In this work, a different approach is advocated, according to which performance optimization is automated. First, this type of automation relieves considerable burden from workflow designers. Second, automated optimization yields intrinsically more flexible and resilient workflows. Increased flexibility stems from the fact that several equivalent alternatives are investigated by the optimizer thus providing more options. Also, when external conditions that impact on the workflow performance evolve, automatically re-optimizing the workflow is important for efficiently adapting to the new setting to attain resilience. Third, performance issues are playing an increasingly important role in modern BPM, which becomes more data- and process-intensive, e.g., in order to cope with big data [2] giving rise to the need for performance optimization. Finally, optimizers for automatically deriving execution details is an integral component in systems

© Springer International Publishing Switzerland 2016
M. Ivanović et al. (Eds.): ADBIS 2016, CCIS 637, pp. 19–28, 2016.
DOI: 10.1007/978-3-319-44066-8_2

that aim to allow end users to submit the process definition at a more declarative level, e.g., as discussed in [7].

Performance optimization is a field that has been largely investigated in databases (e.g., [5]) and data-centric flows (e.g., [6,9]). Although these techniques cannot solve the problem of business process optimization in its entirety, they can form a starting point for automated performance optimization, as explained in this work. The key first step is to bridge the modeling gap: data-centric flows are typically represented as directed acyclic graphs (DAGs) and optimization techniques rely on statistical metadata, such as cost per task invocation and selectivity, which can be regarded as annotations to these DAGs. We adopt the same modeling abstraction for business processes, and we explain how BPMNv2.0 elements are translated to such annotated DAGs. The intention is to keep using the BPMN standard and the mapping to a DAG, along with the subsequent optimizations, to occur in a way transparent to the process designer automatically. In summary, the main contributions of this paper is the introduction of an annotated DAG-based approach to BPMNv2.0 modeling along with concrete examples of mappings of the main BPMNv2.0 elements, and the presentation of the optimizations enabled.[1]

In the remainder of this section, a motivating example and an overview of related work is provided. Next, we elaborate on our DAG modeling abstraction. In Sect. 3, the proposed mappings of the main BPMNv2.0 elements are provided. The optimization opportunities and the open research issues are discussed in Sects. 4 and 5, respectively.

Motivating Example. A sub-process that is encountered in banks for processing loan requests is as follows. Upon receiving a customer application, an employee fills in the applicants personal details, and then performs a series of tasks contacting trusted services from third parties on the web. Such tasks include the following: to import additional customer personal data, to check if the applicant is on any black list, to check the borrowing capacity and to check the information with the help of the national credit bureau. If any of these checks fails, the process aborts and the application is rejected. Finally, the third party services are invoked by providing the customer's SSN identity number.

This scenario is simple but capable of showing a set of optimization issues involved. We give some examples: *Which is the optimal sequence to contact the third-party services in terms of performance, given that several orderings are valid (e.g., it does not matter whether the check of the borrowing capacity precedes the check regarding the black lists and vice versa)? Should the invocations be performed in parallel? Should an employee fill full personal details only after the checks have passed?* In the envisaged approach, one can take these decisions automatically, in a principled manner in the sense that cost-based algorithms (which may well be accompanied by theoretical optimality guarantees) can be employed. Further discussion on this is deferred to Sect. 4.

---

[1] A more extended version of this work is in [3].

Related Work. Automatically devising executable workflows that speed-up execution or improve on other performance metrics is an overlooked area in business process management (BPM). Performance optimization is considered in the context of *process redesign*, which covers several topics, as discussed in [1]. Some examples are to divide an existing process into two or more separated processes, to eliminate obsolete activities, to assign tasks to the more specialized person and, in general, to perform judicious responsibility assignment, and to buffer requests to external information sources. The most relevant heuristics to database-like optimization are the so-called "business process behavior heuristics", which include *re-sequencing, knock-out*, and *parallelism*. Re-sequencing covers the optimizations that involve changing the execution order of activities, while preserving the process semantics and correctness. A specific form of re-sequencing is to move activities that check conditions, which if not met, lead to process termination, as early as possible. Such activities are termed as knock-out ones. Parallelism deals with decisions as to whether some activities should be executed in a sequential or a time-overlapping fashion. We target exactly these form of optimizations, but in a cost-based manner instead of using ad-hoc heuristics. Finally, there are techniques that restrict their optimizations in the data management tasks within business processes, e.g., [9].

2   The Proposed DAG-Modeled Abstraction

In data-centric flows, which are also described as DAGs, each graph vertex corresponds to a task. The tasks manipulate data (e.g., extract sentiment information from tweets, combine user identifier numbers with customer info from an underlying database, and so on), and the edges denote how the transformed data flow across the tasks. Since performance in these data-centric flows is directly dependent on the volume of data being processed and the capacity of the execution engines, the optimization methodologies aim to process as fewer data as possible and make judicious assignment of tasks to resources. For the former, the key idea is to prune unnecessary data, that is data that do not contribute to the flow final desired result, as early as possible.

In business processes, the things that flow across tasks are "tokens". So, our DAGs emphasize the volume of the tokens flowing rather than the business logic and the control of the flow. Each BPMN task corresponds to a vertex in the DAG. A directed edge connects each ordered pair of vertices, between which a transmission of tokens takes place in the context of the process.

The goal of the performance optimization is to improve the average performance across multiple process executions. Performance can be crisply defined in several ways, e.g., in Sect. 4 we sum the costs of each activity, which reflects how efficiently tokens are processed. Statistical metadata drive the optimization procedure. This metadata are typically extracted from log files. The exact type of metadata depend on the specific optimization problem, but two types are most commonly encountered: *selectivity* and *cost* (per invocation). Selectivity is the average ratio of output to input tokens. For example, a task that, for each given recipe, triggers a task to prepare a single meal has selectivity 1. If a task performs a check and may cause early process termination in case of the test failure, the selectivity is lower than one. Analogously, if a task produces multiple tokens per input token on average, its selectivity is above one. In all the cases, the output tokens flow across all outgoing edges of the corresponding vertex. The cost of a task is measured in the same units as the performance criterion. If performance is measured in time units, then the task cost is the average time needed to execute that task. In BPMN processes, things (captured by tasks) have to be done under certain circumstances (captured by gateways); in our DAG-based approach, the statistical metadata play an important role in considering the gateways semantics through annotations to vertices.

Apart from the statistical metadata, there are two other categories of metadata needed. The first category covers dependency constraints between task pairs, i.e., whether a task must precede another one in any execution plan to preserve semantic correctness or whether two tasks must be placed on distinct DAG branches, where each DAG branch corresponds to a different execution path. The second category captures behavioral characteristics, such as whether the task can be parallelized, so that its workload is executed by multiple executors in parallel, and whether a task can operate in a pipelined manner, i.e., to be capable of producing output tokens before consuming its entire input.

A distinctive feature of our proposal is that BPMN tasks correspond to DAG vertices but the opposite does not necessarily hold. We employ the notion of artificial tasks termed as *dummy* tasks. Overall, the combination of normal and dummy tasks with appropriately set statistical metadata allow for modeling the token flow in business processes and paves the way for performance optimization, as discussed next.

3   BPMNv2.0 Symbol Mapping

In this section, we describe how we can model the main elements of BPMNv2.0 to our annotated DAGs. The examples are deliberately simple, to convey easier our message, and they are drawn from the Camunda platform.[2] To avoid confusion, the tasks in our DAGs are depicted as circles rather than rounded rectangles. We explicitly discuss activities and gateways; [3] includes also event discussion.

Task. An ordinary task, independently of its exact type (e.g., manual, service, business rule, and so on), is mapped to a distinct vertex in our model. The cost metadata is the average cost in time units to execute that task, and its selectivity is the average ratio between the output and input tokens.

For example, in Fig. 1, there is the task *"prepare meal"*. This task is triggered after it has received the menu suggestions from the previous task. If the average time to prepare a meal is *c_pm*, then the cost of that vertex takes that value. The selectivity is set to 1, because, for each suggestion, there is a single meal prepared.

---

[2] http://camunda.org/bpmn/reference/.

The loop tasks, like the "*suggest dish*" one in the figure, require a bit more attention, because they execute more than one time on average. Let us suppose that the average number of times the task is activated is n and the average time cost to execute each time is c_sd. There are two ways to handle this

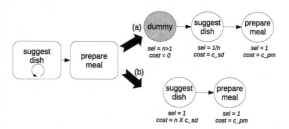

Fig. 1. Mapping a loop and an ordinary task.

case. First, we can insert a zero cost *dummy* task before "*suggest dish*". The selectivity of the dummy task is set to n, whereas the cost of the vertex corresponding to "*suggest dish*" remains c_sd. However, the selectivity of this vertex needs to become $1/n$ to account for the fact that even if n times the "*suggest dish*" task is executed, there is always one token passed on to the subsequent task. The second option is not to use a dummy task and amortize the cost of the vertex, so that it captures the fact that, on average, it is executed n times and thus becomes $n \times c\_sd$. Both options are shown in Fig. 1.

In the previous examples both tasks are not parallelizable. The multiple instance tasks can be modeled in the same way as the loop ones, but the difference is that they can be parallelized up to a parallelism degree of n.

Compensation Tasks. Compensation tasks can be mapped to our DAG with the help of dummy tasks as well. Consider the example in Fig. 2, where a *book trip* task with cost c_bt is associated with a compensating task *cancel trip* with cost c_ct. Let as also assume that the probability of not triggering the

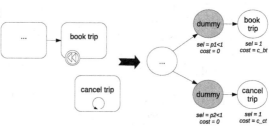

Fig. 2. Mapping a compensation task.

compensating task and continuing the normal execution is $p1$, whereas the probability of canceling the trip is $p2 = 1\text{-}p1$. The mapping to our DAG involves two dummy tasks, which do not contribute to the cost, but control the amount of flow to each of the two branches in a way proportional to the afore-mentioned probabilities through setting their selectivities accordingly. The preceding task in this example sends its output to both branches in line with the edge interpretation in our DAG model, and it is the responsibility of the dummy tasks to perform the filtering. The dependency constraints state that none of the two initial tasks should precede the other, i.e., the optimizer cannot place them in a sequence. Note that a simpler mapping would also be possible in cases where the two branches merge just after the book and cancel tasks. In that mapping, we could omit the dummy tasks and have only the *book trip* task with selectivity set to 1 and a weighted cost equal to $p1 \times c\_bk + p2 \times c\_ct$. This mapping does not capture the complete business logic, but is adequate for performance optimization. Similarly, if there are subsequent tasks following *book trip* but no output edge for *cancel trip*, we could have a single task with the weighted cost as above and the selectivity being equal to $p1$.

Subprocess and Call Activities. Subprocesses do not pose any specific challenge per se with regards to their mapping. However, for optimization purposes, it is always more desirable to expand them in order to broaden the optimization search space of the algorithms, provided that those algorithms are capable of navigating efficiently through the expanded search space. Also, from the performance point of view, call activities can be treated like ordinary activities in the way described above.

Adhoc. Adhoc subprocesses contain several tasks that can be executed at any order. This is exactly the sweet spot for database-like optimization, which can decide on the optimal order in a principled manner. In Fig. 3(top), we present an example with an adhoc subprocess with 4 tasks that can be executed in arbitrary order. We map them in the way shown in the right part of the figure; all the tasks in

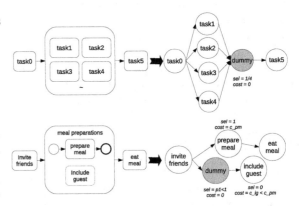

Fig. 3. Mapping an ad-hoc task (top) and an event subprocess (bottom).

the adhoc subprocess are directly connected to the preceding task to denote that there are no inter-dependencies among them and, as such, can be computed in parallel (although the final decision rests with the optimizer as discussed later). Also, we use a zero-cost dummy combiner task to aggregate the output tokens of the adhoc tasks and call the next activity. The selectivity of that dummy task is set to 0.25 because it outputs one token for every four tokens received as input. Further, it is not pipelining, because it needs to consume all its four input tokens in each execution, before creating an output token.

Event Subprocess. An event subprocess may be executed while the enclosing subprocess is active. An example is presented in Fig. 3(bottom), where the enclosing process is a task *prepare meal*, during which new guests can be included (captured by the event task *include guest*). The costs of these tasks are c_pm and c_ig, respectively. To map this case to our token-flow DAG, we insert a dummy filtering vertex with selectivity equal to the probability of executing the event subprocess *p1*. Note that, by definition, *c_ig* is always less than *c_pm*, and the two activities cannot be executed sequentially.

Exclusive, Parallel and Inclusive Gateways. Gateways is a core BPMNv2.0 element and a distinctive feature of process-centric flows not appearing in data-centric workflows. As such, their effective mapping is of high significance in our approach. We distinguish between exclusive, parallel and inclusive gateways. An example of an exclusive gateway is in Fig. 4, where there is an option during meal preparation for the three dishes shown. Let us assume that the average probability to select each of the three options is *p1*, *p2*, and *p3*, respectively; these probabilities need to sum to 1 to account for the fact that always exactly one option is selected. Then, we can insert zero cost dummy filtering tasks with their selectivity set equal to the probabilities above. There is also a zero-cost unary-selectivity dummy combiner task being responsible for synchronization, but this is optional. To preserve the gateway semantics, the dependencies between these vertices are that they cannot be placed in a sequence.

Now, suppose that, in the previous example, the gateway is transformed to a parallel one. Then the dummy tasks on the left become optional (corresponding to zero cost and selectivity equal to 1), so that the three previous options can be directly connected to *choose recipe*. However, the dummy task on the right becomes compulsory and its selectivity is set to 1/3, derived by a generic formula: ratio of 1 to the number of tasks after the parallel gateway. This is to ensure that *eat meal* receives a single token for each *choose recipe* exe-

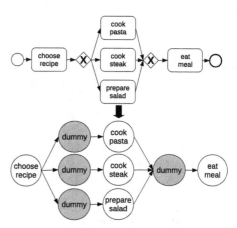

Fig. 4. Mapping an exclusive gateway.

cution no matter how many intermediate tasks are executed in parallel. In addition, the dummy task on the right enforces synchronization.

Inclusive gateways allow tokens to flow across one, many or all paths, i.e., the combine certain features from the exclusive and parallel cases. The high-level mapping shown in Fig. 4 holds for inclusive gateways as well, but with all dummy tasks being compulsory. Contrary to the case of exclusive gateways, the selectivities of the dummy tasks preceding the options can sum to a value greater than 1. In addition, the dummy combiner task on the right part becomes compulsory (as in parallel gateways) and its selectivity is set to the ratio of 1 to the sum of the selectivities on the left.

Gateways and Loops. Gateways are very common to be accompanied by cycles in business graph models, as shown at the top of Fig. 5. We can combine the approaches presented earlier regarding loops and gateways in order to render our graph acyclic. The tasks belonging to the loop path are placed in a sequence with their cost having been amortized as shown in Fig. 1; the alternative of a dummy task in that figure is also valid. Then, the rest of the tasks after the gateway are treated as in Fig. 4.

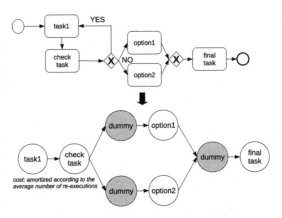

Fig. 5. Mapping a more complex exclusive gateway.

Event-Based vs. Data-Based Gateways. In BPMNv2.0, there is a distinction between data-based and event-based gateways. The former choose the routing of a token to one or more paths according to data associated with the specific token. The latter make decisions based on events happening. From the performance point of view, there is no essential difference between these two cases. For example, consider the exclusive case. Instead of monitoring the probabilities of task activation, in an event-based exclusive gateway, we can monitor the probability of corresponding events and set the selectivities in our DAG vertices accordingly. Next, we discuss BPMNv2.0 events in detail.

4   Optimization Opportunities

The optimizer envisaged receives an initial mapping, and the set of statistical and dependency metadata and derives an optimal execution plan. A key aspect is to decide the exact order of task execution in a cost-based manner.

To provide insights into the benefits, we extend the motivating example, where the ordering of some tasks is flexible thus generalizing adhoc tasks in BPMN. Suppose a simple process, which contains n activities forming a chain (i.e., there are no branches). If, due to dependency constraints, there is no flexibility in the order, then this implies that there are $\frac{n(n-1)}{2}$ dependency edges, either explicitly stated or implied through transitive closure. For example, a loan pre-processing template may define the order in which the tasks for importing contact information of the applicant, checking the borrowing capacity and contacting the credit bureau take place, despite the fact that any ordering is valid.

Figure 6 shows the performance improvements after simulating 100 randomly generated DAGs, where there are $0.75\frac{n(n-1)}{2}$ and $0.5\frac{n(n-1)}{2}$ dependency edges, n ranges from 10 to 15, the selectivity of tasks ranges from 0.01 (extremely filtering) to 2, and the cost ranges from 1 to 100. The value distribution is uniform. The exact optimization metric selected is the sum of the average execution time for each task. As baseline performance, which corresponds to normalized value 1, we consider the running time of the initial DAG before re-orderings.

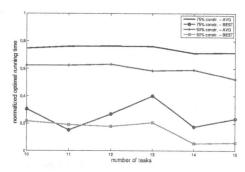

Fig. 6. Example benefits when optimally reordering activities.

In the figure, we can see that, on average, there is a reduction in the running time by 25.62 % for the more constrained case; moreover, the average reduction becomes 40.37 %, for 50 % constraints. Also, there are isolated runs, where the improvements can be up to an order of magnitude, as shown by the maximum improvement plots at the bottom of the figure. These numbers indicate how significant the performance benefits can be, even in simple processes. Exploring all the orderings, even in highly constrained settings, is an intractable problem. The techniques in [6] show how the optimizer can navigate through the search space in a scalable manner. Other performance optimization problems can be considered as well, as discussed in [3].

5 Main Research Issues and Conclusions

Here, we mention the main research issues for developing complete solutions for performance optimization in BPMN business processes.

- *Need for dependency-aware optimization algorithms.* Mapping BPMN models to our DAG abstraction is a necessary but not sufficient condition to perform cost-based performance optimization. In the previous section, we referred to several algorithmic techniques that consider only precedence constraints, i.e., constraints of the form that task A must precede task B. This type of constraints needs to be complemented by (i) parallelism constraints that enforce tasks to be placed in different execution paths; (ii) blocking vs. pipelining information for each task; and (iii) parallelism capability information, to define which tasks are amenable to parallel execution and up to which degree of parallelism. More research is needed to develop solutions that account for the complete range of constraints in business processes.
- *Statistical Metadata Collection.* The statistical metadata play a crucial role, and their efficient collection requires special attention. Techniques like [4] may act as a starting point. Challenges include the fact that statistics are actually

correlated, which means that changing the order of tasks may affect their statistical metadata.

- *Extensive Evaluation.* There needs to be extensive evaluation using benchmarks to reason with confidence about the actual capability of each proposed optimization technique.
- *Mapping to BPMN models and end-to-end solutions.* Holistic solutions should involve the mapping of the optimized execution plan back to a BPMN model and, ideally, be exposed as a software plugin to existing platforms rendering the optimization fully transparent to the process designer.

Finally, it should be remarked that, in BPM, optimization for performance only is inadequate; the focus should also be shifted to aspects such as fault-tolerance, reliability, economic cost and so on.

Summary. This work is motivated by the fact that currently, performance optimization of business process is a manual activity in the responsibility of the designer. To address this limitation, automated performance optimizations should be applied. We explain how we can build upon the knowledge in the data management community to optimize data-intensive queries and flows. More specifically, we discuss the annotated DAG modeling abstraction required to employ such solutions, going through the handling of the main BPMNv2.0 elements in detail. We provided insights into the potential performance benefits and identified the main research issues for enabling automated optimization in BPM.

References

1. Dumas, M., Rosa, M.L., Mendling, J., Reijers, H.A.: Fundamentals of Business Process Management. Springer, Heidelberg (2013)
2. Gao, X.: Towards the next generation intelligent BPM – in the era of big data. In: Daniel, F., Wang, J., Weber, B. (eds.) BPM 2013. LNCS, vol. 8094, pp. 4–9. Springer, Heidelberg (2013)
3. Gounaris, A.: Towards automated performance optimization of BPMN business processes. CoRR abs/1508.07455 (2015)
4. Halasipuram, R., Deshpande, P.M., Padmanabhan, S.: Determining essential statistics for cost based optimization of an etl workflow. In: EDBT, pp. 307–318 (2014)
5. Ioannidis, Y.E.: Query optimization. ACM Comput. Surv. **28**(1), 121–123 (1996)
6. Kougka, G., Gounaris, A.: Optimization of data-intensive flows: Is it needed? is it solved?. In: DOLAP, pp. 95–98 (2014)
7. Lamprecht, A.L., Naujokat, S., Schaefer, I.: Variability management beyond feature models. IEEE Comput. **46**(11), 48–54 (2013)
8. Stiehl, V.: Process-Driven Applications with BPMN. Springer, Switzerland (2014)
9. Vrhovnik, M., Schwarz, H., Suhre, O., Mitschang, B., Markl, V., Maier, A., Kraft, T.: An approach to optimize data processing in business processes. In: VLDB, pp. 615–626 (2007)

Pixel-Based Analysis of Information Dashboard Attributes

Jiří Hynek[1]([✉]) and Tomáš Hruška[2]

[1] Department of Information Systems, Faculty of Information Technology,
Brno University of Technology, Božetěchova 1/2, Brno, Czech Republic
ihynek@fit.vut.cz
[2] IT4Innovations Centre of Excellence, Faculty of Information Technology,
Brno University of Technology, Božetěchova 1/2, Brno, Czech Republic
hruska@fit.vut.cz

Abstract. This paper focuses on pixel-based usability guidelines and their use for an information dashboard user interface. The first part of the paper examines existing usability design advices, presents existing pixel-based metrics and make suggestions of new ones. The second part presents results of pixel-based analyses performed on two groups of well-designed dashboards and randomly chosen dashboards. Results of these two groups are compared and their differences are discussed.

Keywords: Information dashboard · Pixel-based analysis · Usability guidelines · User interface

1 Introduction

An information dashboard, as a presentation layer of an information system, is a tool whose goal is to visualize concrete information which is important for an accomplishment of particular tasks. It usually presents analytical data and key performance indicators of some processes which are used for further decisions. An information dashboard (*dashboard*) should help make these decisions. Thus, it should be well-designed to help users get quickly familiarized with its meaning and displayed content. Visual aspect of a dashboard should be in accordance with design principles of user interfaces used for a data visualization.

As described in [3], most of the existing dashboards contain some design problems. In some cases, these problems represent little aesthetic imperfections, but sometimes, these problems may lead to serious usability problems. Possible way how to detect these problems is to let users or design experts perform this task (e.g. *usability testing* or *heuristic evaluation* [13]). It can be effective but expensive and time consuming. In our research, we are working on system which can do these usability evaluation automatically without presence of users or design experts. For this purpose we designed framework described in [5].

The goal of this paper is to collect pixel-based metrics which can be used for computing attributes suitable for differentiation between well designed dashboards (according to advices described in [3]) and the others. We expect that

© Springer International Publishing Switzerland 2016
M. Ivanović et al. (Eds.): ADBIS 2016, CCIS 637, pp. 29–36, 2016.
DOI: 10.1007/978-3-319-44066-8_3

correctly designed dashboards can be numerically distinguished by specific values of particular attributes. We explore this assumption by comparisons of computed attributes for well design dashboards and randomly picked ones.

2   Usability Evaluation Based on Guidelines

A definition of a dashboard was established by Stephen Few [3] as:

Definition 1. *A dashboard is a visual display of the most important information needed to achieve one or more objectives; consolidated and arranged on a single screen so the information can be monitored at a glance.*

To design such dashboard, [3] recommends several advices and demonstrates frequently made mistakes. Example of such advice can be seen in Fig. 1. Similar advices can be found in a lot of literature [4, 8, 13, 18, 20].

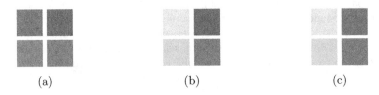

      (a)   (b)   (c)

Fig. 1. Example of the design principle which recommends to encode a different meaning by color intensity (b, c) rather than only by color hue (a) because it can be missed by someone who is color-blind.

In this research, we focus on quantitative advices (*usability guidelines*) which can numerically measure some values and thus they can be simply transformed into a runnable code. They are usually simpler than qualitative advices. Therefore, lower rate of usability problems and higher rate of false positive usability problems are detected. However, they are not intended to replace evaluation based on qualitative advices (e.g. heuristic evaluation). Their goal is to automatize some time-consuming processes.

The role of usability guidelines has been rising with the evolution of user interfaces. In the 80s, guidelines was designed for textual user interfaces [17]. In the 90s, guidelines were integrated in several systems used for graphical user interfaces development – e.g. [2, 9, 11], which are also evaluated by [6]. In the early 2000s, the evolution of the Internet and web pages played crucial role in a development of a usability evaluation [7]. Researchers were looking for new techniques which would increase attractiveness of web pages. Terms like an *aesthetics* which specifies a rate of webpage beauty became more important [10]. Several guidelines suitable for increasing of aesthetics were found [12, 14–16, 22]. Today, smartphones, tablet devices and *internet of things* are becoming more important. Thus, further progress of user interface design regarding these areas could be expected.

Following list presents selected guidelines based on color pixel-based metrics which measure usage of individual color values, or a distribution of those values in a dashboard image raster:

- **Colorfulness:** This attribute represents a diversity of used colors. According to [3], it should be low in dashboard. To monitor such attribute, dashboard raster representation need to be converted into color space which better corresponds with a human perception (like *HSB* [3] or *CIE Lab* [20,22]). For instance, [16,21] consideres colorfulness as a saturation in CIE Lab color space where saturation is computed as *chroma* divided by *lightness.*
- **Amount and share of color values:** According to [3], a dasboard should contain low amount of dominant color values. To measure such attributes, raster can be transformed to RGB color space. Color values of this color space are usually stored as 24 bit numbers (more than 16.77 million of distinct color values), thus *posterization* [15] or conversions to 8 bit Gray Scale (representing color intensity) and 1 bit Black-White (*thresholding* [1]) color spaces are advised.
- **Distribution of colors:** We focused on two design attributes – *balance* and *symmetry*, illustrated in Fig. 2. Balance is a metric which calculates a distribution of an optical weight in a picture along a vertical or horizontal axis [14,19]. Symmetry is a metric, which calculates a rate of axial duplication of a visual image of graphical elements along horizontal and vertical axes (*axial symmetry*) or the diagonal axes (*radial symmetry*) [14,22]. Higher balance and symmetry can make user interface less disordered which can lead to simpler perception of these interfaces. To compute these attributes, we used formulas presented by [9,14] and extended them for pixel-based purposes (color intensity of pixels was also considered – Fig. 2c).

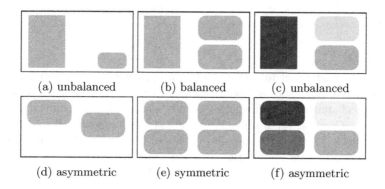

(a) unbalanced   (b) balanced   (c) unbalanced

(d) asymmetric   (e) symmetric   (f) asymmetric

Fig. 2. Example of balanced and unbalanced (symmetric and asymmetric) screens. Balance and symmetry based on color intensity (c, f) has been also considered in this research.

3   Experiment Description

The goal of our experiment was to find out, if there exist metrics which can be used to distinguish between a group of dashboards designed according to expert advices and a group of dashboards which were designed without these rules. We applied selected metrics on the two following groups:

- **Group 1:** a group of 10 dashboards which were designed by experts according to rules defined by [3].
- **Group 2:** a group of 120 randomly picked dashboards which were collected from the Internet. No information about usability of these dashboards was considered.

Each dashboard was stored as a bitmap in 24 bit RGB color space. For purposes of particular analyses, further transformations to other color spaces were done. The dashboards of Group 1 were transformed to other 3 sizes (75 %, 66 %, 50 % of the original size[1]) to take a lower resolution impact into consideration. Therefore, the size of Group 1 was increased from 10 to 40 samples.

Table 1. The list of metrics used in each analyzed dashboard (B&W = Black-and-White)

Metric/Set of metrics	Color space
colorfulness: average hue, saturation, brightness	HSB
colorfulness: average lightness, chroma, hue, saturation	CIE Lab, CIE Lch
amount of distinct color values	RGB, Gray-Scale
... with share higher than 0.1 %, 0, 5 %, 1 %, 5 %, 10 %	Gray-Scale
share of the 1st and 1st+2nd most used color values	RGB, Gray-Scale
share of black color value	B&W
balance, symmetry	Gray-Scale, B&W

The experiment procedure was the following. First, we computed values of a selected metric for each dashboard of Group 1 and 2. Then, we took these values and calculated an arithmetic mean μ and a standard deviation σ for Group 1 and Group 2 separately. We repeated this procedure for each metric listed in Table 1. Finally, we compared results of Group 1 with results of Group 2. We were also looking for metrics with low standard deviation in Group 1 to find typical characteristics of well-designed dashboards.

---

[1] The smallest width was 175 px and the smallest height was 130 px. There was no reason to include dashboards with smaller resolution, because they were readable with difficulties. The experiment was focused only on dashboard with sufficient resolution.

4   Results

First dashboard attribute, that was examined, was colorfulness. We used formula based on the sum of mean saturation and its standard deviation described by [21]. As we expected, the dashboards of Group 1 are less colorful than those ones from Group 2 (see Table 2). This is satisfied for both examined color spaces – HSB a CIE Lab. Thus, these metrics seems to be suitable for further application in dashboard categorizations. As described in Sect. 3, we also applied the same formula with other channels of mentioned color spaces (like hue or brightness), yet the results were not so interesting.

We also did an experiment and converted dashboards to 8 bit Gray-Scale color space and analyzed histograms of color intensities. As it can be seen in Fig. 3, the histogram of the dashboard with a low rate of colorfulness (Fig. 3a) contain one dominant intensity (background) and few other intensities with a low frequency of occurrence (data pixels). On the contrary, the histogram of the colorful dashboard (Fig. 3b) contains a range of many color intensities with a relatively high frequency of occurrence. That makes the share of the dominant color intensity significantly lower.

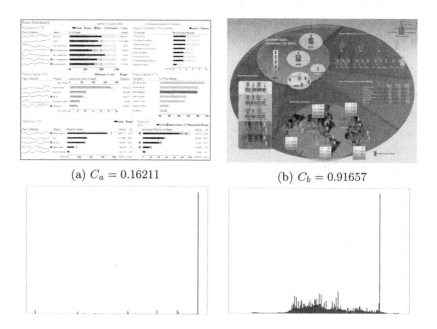

(a) $C_a = 0.16211$ (b) $C_b = 0.91657$

Fig. 3. Comparison of 2 Sales Dashboards used from [3] with histograms of their color intensities – from 0 (black) to 255 (white) (Labels presenting values of vertical axis are ignored. They are not so important in this case because they depend on actual size of a dashboard and our intention was to emphasize ratios of color intensities.). The dashboard on the left is significantly less colorful than the one on the right side.

The reason of this difference is also the fact that dashboards of Group 2 often contain high amount of *non-data pixels* [18] whereas amount of these pixels is highly reduced in Group 1. This reduction causes increasing of background share which is represented by the most used color value in Group 1. The best color spaces to numerically evaluate share of major colors are according to our results 12 bit posterized RGB, 4 bit Gray-Scale and 1 bit thresholded Black&White (see Table 2). Thresholding was done adaptively according to [1].

Table 2. Selection of the most interesting results which were gathered by application of measures on 2 groups of dashboards described in Sect. 3.

Metric	μ_1	σ_1	μ_2	σ_2
HSB, CIE Lab/Lch				
HSB saturation	0.1094	0.0459	0.3841	0.2081
CIE Lab/Lch saturation	0.2170	0.1823	0.6835	0.5770
12 bit RGB ($2^{12} = 4096$ color values)				
Amount of distinct color values	375	435	670	464
Share of the 1st most used color value	79.94 %	7.67 %	54.44 %	21.08 %
Share of the 1st+2nd most used color values	84.60 %	4.71 %	66.37 %	18.95 %
4 bit Gray-Scale ($2^4 = 16$ color values)				
Amount of colors with share > 5 %	1.35	0.53	3.36	1.65
Amount of colors with share > 10 %	1.00	0.00	2.06	0.99
Share of the 1st most used color value	82.94 %	4.87 %	57.06 %	19.57 %
Share of the 1st+2nd most used color value	87.10 %	3.51 %	72.54 %	17.96 %
Color distribution: balance	0.914	0.041	0.764	0.138
Color distribution: symmetry	0.921	0.021	0.852	0.061
1 bit Black-and-White (2 color values)				
Share of black color value	13.98 %	3.33 %	28.80 %	15.77 %

As regards color distribution analysis, dashboards of both examined groups have high rate of balance and also relatively high rate of symmetry in the all three color spaces. The best results were computed in 4 bit Gray Scale color space (presented in Table 2). It can be seen that balance and symmetry values are slightly higher in Group 1, which was expected. However, the difference between Group 1 and 2 is not so significant. We assume that this fact is caused by human need to see things balanced and symmetric and thus designers usually provide these attributes without an explicit intention. These metrics seem to be good to evaluate dashboard usability but they don't seem to be the best at differentiation between Groups 1 and 2.

5   Summary

The goal of this research was to analyze pixel-based attributes of dashboards. For this purpose we specified the set of the measures which were used in 3 particular analyses examining colorfulness, color usage and color distribution. We analyzed the group of well-designed dashboards and compared the results with the set of randomly chosen dashboards with no explicit information about their usability. As result, we identified metrics which are suitable for classification algorithms which will be able to distinguish well-designed dashboards (Table 2).

5.1   Limitations and Future Work

In this research, only some of pixel-based attributes were examined. In the future, we would like to add attributes which relate with representation based on graphical elements. Also, we would like to apply our framework in some real design tool which is used for dashboard creation. For now, our training set is based on dashboards recommended by [3]. In the future, we would like to use proposed metrics to calculate dashboard attributes which will be used as features in machine learning algorithms. Then, other training sets based on other design principles could be used.

Acknowledgment. This work was supported by The Ministry of Education, Youth and Sports from the National Programme of Sustainability (NPU II) project "IT4Innovations excellence in science – LQ1602".

References

1. Bradley, D., Roth, G.: Adaptive thresholding using the integral image. J. Graph. GPU Game Tools **12**(2), 13–21 (2007)
2. Bodart, F., et al.: Towards a dynamic strategy for computer-aided visual placement. In: Proceedings of the Workshop on Advanced Visual Interfaces, pp. 78–87. ACM (1994)
3. Few, S.: Information Dashboard Design. O'Reilly, Cambridge (2006)
4. Gibson, J.J.: The Perception of the Visual World. The Riverside Press, Cambridge (1950)
5. Hynek, J., Hruška, T.: Automatic evaluation of information dashboard usability. Int. J. Adv. Comput. Sci. Appl. (IJCSIA) **5**(2), 383–387 (2015). (IRED)
6. Ivory, M.Y.: An empirical foundation for automated web interface evaluation. Doctoral dissertation, University of California at Berkeley (2001)
7. Ivory, M.Y., Hearst, M.A.: The state of the art in automating usability evaluation of user interfaces. ACM Comput. Surv. (CSUR) **33**(4), 470–516 (2001)
8. Johnson, J.: Designing with the Mind in Mind: Simple Guide to Understanding User Interface Design Guidelines. Elsevier, Amsterdam (2013)
9. Kim, W.C., Foley, J.D.: Providing high-level control and expert assistance in the user interface presentation design. In: Proceedings of the INTERACT 1993 and CHI 1993 Conference on Human Factors in Computing Systems, pp. 430–437. ACM (1993)

10. Lavie, T., Tractinsky, N.: Assessing dimensions of perceived visual aesthetics of web sites. Int. J. Hum. Comput. Stud. **60**(3), 269–298 (2004)
11. Mahajan, R., Shneiderman, B.: Visual and textual consistency checking tools for graphical user interfaces. IEEE Trans. Softw. Eng. **23**(11), 722–735 (1997)
12. Moshagen, M., Thielsch, M.T.: Facets of visual aesthetics. Int. J. Hum. Comput. Stud. **68**(10), 689–709 (2010)
13. Nielsen, J.: Usability Engineering. Elsevier, Amsterdam (1994)
14. Ngo, D.C.L., et al.: Modelling interface aesthetics. Inf. Sci. **152**, 25–46 (2003)
15. Purchase, H.C., Freeman, E., Hamer, J.: An exploration of visual complexity. In: Cox, P., Plimmer, B., Rodgers, P. (eds.) Diagrams 2012. LNCS, vol. 7352, pp. 200–213. Springer, Heidelberg (2012)
16. Reinecke, K., et al.: Predicting users' first impressions of website aesthetics with a quantification of perceived visual complexity and colorfulness. In: Proceedings of the SIGCHI Conference on Human Factors in Computing Systems, pp. 2049–2058. ACM (2013)
17. Smith, S.L., Mosier, J.N.: Guidelines for designing user interface software. Mitre Corporation (1986)
18. Tufte, E.R.: The Visual Display of Quantitative Information, 2nd edn. Graphics Press, USA (2001)
19. Vanderdonckt, J., Gillo, X.: Visual techniques for traditional and multimedia layouts. In: Proceedings of the Workshop on Advanced Visual Interfaces, pp. 95–104. ACM (1994)
20. Ware, C.: Information Visualization: Perception for Design. Morgan Kaufmann Publishers, San Francisco (2004)
21. Yendrikhovskij, S.N., et al.: Optimizing color reproduction of natural images. In: Color and Imaging Conference, vol. 1998(1), pp. 140–145. Society for Imaging Science and Technology (1998)
22. Zheng, X.S., et al.: Correlating low-level image statistics with users-rapid aesthetic and affective judgments of web pages. In: Proceedings of the SIGCHI Conference on Human Factors in Computing Systems, pp. 1–10. ACM (2009)

Towards Adaptive Distributed Top-*k* Query Processing

Claus Dabringer and Johann Eder$^{(\boxtimes)}$

Department of Informatics Systems,
Alpe Adria University Klagenfurt, Klagenfurt, Austria
{Claus.Dabringer,Johann.Eder}@aau.at

Abstract. ADiT is an adaptive approach for processing distributed top-*k* queries over peer-to-peer networks optimizing both system load and query response time. It considers the size of the peer to peer network, the amount *k* of searched objects, and for each peer: the bandwidth, the amount of objects stored, and the speed of in processing a local top-*k* query. In extensive experiments with a variety of scenarios we could show that ADiT outperforms state-of-the-art distributed query processing techniques.

Keywords: Distributed query processing · Top-*k* query · Peer-to-peer databases · Federated databases

1 Introduction

Top-*k* queries retrieve the *k* tuples of a query result which score best for a given objective function. Top-*k* queries help to overcome the problem of too large query results on one hand and too low recall, if the query is more constrained, and are therefore a promising technology for improving and accelerating search in for various data collections, e.g. for the search for suitable samples in biobanks [10], our main application area. Top-*k* queries are also popular for providing users ranked search results they are used from web search engines. Top-*k* queries, in particular, the optimization of top-*k* query processing for central databases received a lot of attention [3,5,8,9,16–20,23]. Optimizing top-*k* queries in distributed environment, in particular in highly distributed networks of federated or peer to peer databases still has significant research needs.

Current distributed top-*k* query processing approaches either focus on reducing the amount of transmitted queries [2,15] or on keeping the amount of transported objects low [4,6,12,21], or to reduce the communication costs [11]. However, both the transmitted objects and messages affect the *system effort* and *query response time* in a peer to peer system. Therefore, we introduce an adaptive distributed top-*k* (short ADiT) query processing approach considering both.

The work reported here was supported by the Austrian Ministry of Science and Research within the program GENAU (project GATIB II) and within the project BBMRI.AT.

© Springer International Publishing Switzerland 2016
M. Ivanović et al. (Eds.): ADBIS 2016, CCIS 637, pp. 37–44, 2016.
DOI: 10.1007/978-3-319-44066-8_4

To the best of our knowledge ADiT is the first approach using the amount of messages *and* the amount of transmitted objects to measure *system effort* and *query response time.*

Processing a top-k query in a p2p network with horizontal partitioning involves sending a top-k query to each peer. The optimization problem now is to determine a proper k_p for each p of the peers, i.e. how many objects should be fetched from which peer. If this k_p is too large, it results in unnecessary computation at the peers' site and unnecessary traffic. If it is too low, it is necessary to send additional queries to the peers.

In our approach the following parameters are used to determine k_p, i.e. how many objects should be fetched from which peer: the size of the peer to peer network, the amount k of searched objects, the network capabilities of a connected peer, i.e. the transmission rate, the amount of objects stored on each peer, and the speed of a peer in processing a local top-k query.

In the following we show the general architecture for processing top-k queries in a p2p environment and derive some heuristics based on the parameters outlined above. We describe the implementation of the ADiT approach and show in an extensive set of experiments the performance gains using this approach. An extended version of this paper is provided in [7].

2   ADiT in General

Adaptive Distributed Top-k query processing (short ADiT) is able to process distributed top-k queries over horizontally partitioned data exactly. ADiT assumes a dyxnamic peer to peer network. Each peer has variable bandwidth capabilities and individual message costs. In contrast to other approaches [21,22] ADiT does not rely on caching techniques. Thus the performance does not depend on stable data or on reoccurring queries.

The aim of ADiT is to achieve a low overall system effort as well as a fast query response time. The first parameter, the overall system effort is defined as the total amount of time the peers needed for (1) sending requests to other peers in the network to obtain further objects, (2) searching objects and (3) transmitting objects. The second parameter is the query response time, the time elapsed between submitting a query and the return of the result. Formulas 1 and 2 defines the system effort and the query response time where $MsgCount_i$ is the total amount of messages sent to peer P_i and n_i is the amount of objects retrieved from peer P_i. We use the following abbreviations throughout of this paper: N is the peer to peer network, Q is the top-k query, R is the queried relation, and P_i is a peer in the p2p system.

$$SystemEffort(N, Q, R) = \Sigma(CommCosts(N.P_i, MsgCount_i) + \quad (1)$$
$$DBCosts(N.P_i, Q, R, n_i) +$$
$$TransCosts(N.P_i, R, n_i))$$

$$QueryAnswerTime(N, Q, R) = max(CommCosts(N.P_i, MsgCount_i), \quad (2)$$
$$DBCosts(N.P_i, Q, R, n_i),$$
$$TransCosts(N.P_i, R, n_i))$$

The unit of system effort as well as of query response time is seconds. Thus it is needed to map the different costs to a time factor. Function 3 defines how sending $MsgCount$ requests to peer P is mapped to a time factor. The amount of incoming messages is multiplied with the constant costs that arise when establishing a connection to peer P. This gives the amount of time that is spent by sending $MsgCount$ messages to peer P.

$$CommCosts(P, MsgCount) = P_{MsgCosts} * MsgCount \quad (3)$$

Function 4 defines how retrieving n objects from relation R of peer P is mapped to a time factor. The transmission costs are influenced by the size of the object in relation R on peer P and by the transmission rate of peer P.

$$TransCosts(P, R, n) = \frac{(P_{R_{ObjectSize}} * n)}{P_{TransRate}} \quad (4)$$

The database costs ($DBCosts(N.P_i, Q, R, n)$) for searching the best n objects in relation R on peer P_i strongly depend on the top-k approach used on peer P_i, performance of the answering peer P_i, and the issued query Q, e.g. on the number of restrictions. ADiT assumes that each peer provides an estimate of the time needed to return the top-k objects for a query with m restrictions on a relation with size N. There is no assumption which procedure a peer uses to process top-k queries.

ADiT works iteratively and calculates a separate fetch size k'_p for each peer in each iteration. Then ADiT broadcasts the query Q *in parallel* and gathers the top-k'_p from each peer p. Then ADiT tries to publish objects and repeats if necessary. There are two major possibilities for tuning: Choosing an appropriate fetch size k'_p for each peer in each iteration and avoiding to contact peers which cannot contribute to the result. For choosing the fetch size there are two extreme cases: $k'_p = 1$ leads to a minimal amount of *transmitted objects* but to a higher amount of *transmitted messages*, while $k'_p = k$ leads to a minimal amount of *transmitted messages* but to a higher amount of *transmitted objects*.

3   Heuristic Fetch Size Calculations

Analyzing a large number of queries varying the influencing factors [7] we developed two heuristics (basic and enhanced) for choosing a good fetch size k'_p for each individual peer p.

Basic Heuristics. The basic heuristics shown in Eq. 5 only uses the amount of relevant peers N_{Size} and the amount of searched objects k to derive a common fetch size f for all peers. The basic heuristics does not assume any particular data distribution. Thus it tries to retrieve an equal amount of objects from each peer. In case k is larger than N_{Size} the basic heuristics equally distributes k among the available peers. Otherwise the basic heuristics calculates the smallest multiple of k which is greater or equal than N_{Size} and equally distributes this amount among the available peers. The *consFactor* is used to increase the fetch size since it is unlikely that each peer will contribute the same number of objects. This increasing is used to fetch more objects and keep the number of iterations small. Our initial experiments showed that a *consFactor* of 2 leads to good results, e.g. few iterations and thus few messages exchanged in the p2p network. If the data is not distributed equally, *consFactor* should be chosen higher.

$$f = min(k, consFactor * \lceil N_{Size}/k \rceil * N_{Size/k}) \tag{5}$$

Enhanced Heuristics. The enhanced heuristics calculates the fetch size k'_p for each peer p *separately*. It uses additional parameters to adjust the fetch size for each peer properly:

- $ObjectsStored_p$: Amount of objects stored on peer p.
- $ObjectsStored_N$: Amount of objects stored in the peer to peer system N, i.e. $sum(ObjectsStored_p)$
- $Speed_p$: Query processing speed of peer p, e.g. a value between 1 and 10 where 1 is the slowest and 10 the fastest speed.
- $maxSpeed_N$: Maximum query processing speed of a peer in the peer to peer system N.
- $TransRate_p$: The transmission rate describing how fast the network connection of a certain peer is. This value is given in MBit per second.
- $maxTransRate_N$: Maximum transmission rate of a peer in the peer to peer system N.

The knowledge gathered during query processing iterations comprises the following parameters:

- $ObjectsRetrieved_p$: Amount of objects of peer p which have already been retrieved, initially 0.
- $ObjectsPublished_p$: Amount of objects of peer p which made it in the top-k answers, initially 0.
- $ObjPub_N$: Amount of objects returned to the user, initially 0.

All these parameters are used to calculate different weights which influence the enhanced heuristics. Applying the basic heuristics to the large test scenarios showed that the proposed fetch size should be treated as a lower limit. Therefore, the enhanced heuristics uses the different weights to *increase the fetch size* determined with the basic heuristics. To accomplish that the enhanced heuristics maps its weights to the interval of [1, 2]. This prevents from fetching fewer objects than the basic heuristics suggested. The enhanced heuristics assumes that all previous iterations can be used to reason about following iterations, e.g. it assumes that peers that contributed more objects in previous iterations will also contribute more objects in the following iterations. This assumption is reflected in weight w_{pF} which is defined in Eq. 6. The more objects a peer published compared to all other peers, the more objects are gathered from this peer *in the next iteration*.

$$w_{pF} = (1 + ObjectsPublished_p/ObjPub_N) \qquad (6)$$

The enhanced heuristics tries to reduce the amount of fetched objects which are not needed. Thus it fetches more objects from peers where the ratio between fetched objects and published objects is high. Equation 7 shows the definition of weight w_{uF}.

$$w_{uF} = (1 + ObjectsPublished_p/ObjectsRetrieved_p) \qquad (7)$$

The enhanced heuristics assumes that peers which store more objects will contribute more to the final answer. Thus it suggests to fetch more objects from larger peers. It uses Eq. 8 to incorporate that fact, namely weight w_{DBF}.

$$w_{DBF} = (1 + ObjectsStored_p/ObjectsStored_N) \qquad (8)$$

Since it is cheap to ask a faster peer for more objects the enhanced heuristics defines w_{Speed} and $w_{TransRate}$. Equation 9 models the fact that more objects should be fetched from peers which are faster in searching their databases.

$$w_{Speed} = (1 + Speed_p/maxSpeed_N) \qquad (9)$$

Equation 10 deals with the transmission of objects. It reflects that more objects should be fetched from peers which have a higher transmission rate.

$$w_{TransRate} = (1 + TransRate_p/maxTransRate_N) \qquad (10)$$

The weights described in Eqs. 6–10 are used by the enhanced heuristics to influence the basic heuristics. The weighted fetch size is determined with the heuristic function shown in Eq. 11.

$$k'_p = min(k - ObjPub_N, \lceil f * w_{pF} * w_{uF} * w_{DBF} * w_{Speed} * w_{TransRate} \rceil) \qquad (11)$$

The upper bound for fetch size k'_p is obviously the amount of missing objects, namely $k - ObjPub_N$. The enhanced heuristics does not fetch more objects than the amount of missing objects from any of the peers in the peer to peer system.

4   Prototype and Experiments

ADiT has been completely implemented in PL-SQL as a set of stored procedures. To compare ADiT against a state of the art distributed top-k query processing technique we also implemented the *algorithm with remainder top-k queries (short ARTO)* [21] in this database layer.

We performed experiments on 2 databases: One filled with randomly generated data, and the other consisting of a single relation containing 68 categorical attributes taken from the *UCI Machine Learning Repository* [1,14] which contains over 2.400.000 entries in this single relation which we distributed among the peers in the network such that the size of the database of each peer varied between 5.000 objects and 500.000 objects.

Within this section we present various diagrams generated from the data produced by the conducted test runs. We primarily focused on the *system effort* caused by a certain query and on the *query response time*. To make precise statements about ADiT and the enhanced heuristics we used the basic heuristics with a *consFactor* of 2 and four other heuristics to compare them to the enhanced heuristics: (1) $k'_p = k$, (2) $k'_p = 1$, (3) $k'_p = \lceil \frac{k}{N} \rceil$, (4) $k'_p = \lfloor \frac{k}{N} \rfloor$, and (5) $k'_p = min(k, 2 * \lceil \frac{N_{Size}}{k} \rceil * \frac{k}{N_{Size}})$

For an easier comparison we show the achieved results in form of two ratios: gain with respect to system effort and for for the query response time. Due to space limitations we can only show one viewgraph. More results and a more detailed discussion is available in [7].

4.1 Discussion of Results

In Fig. 1 we can see the $Ratio_{QAT}$ for a query with 4 restrictions. Additionally, these first figures already show that the heuristics $k'_p = 1$ is not a good choice since it involves high interaction between the query initiator and the other peers. We can also observe that for the query response time the gain over ARTO is rapidly increasing when the amount of searched objects increases. The ratio is growing fast because ARTO needs more sequential message processing when the search amount increases (when the first parallel call was not sufficient).

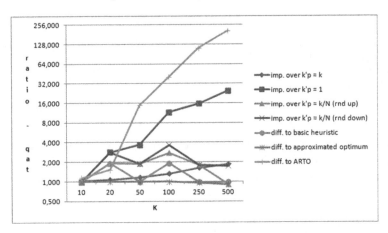

Fig. 1. Ratio for query response time between *enhanced heuristics*, approximated optimum, ARTO and five different approaches to determine the fetch size k'_p in a peer to peer system with 19 peers and varying search amount K and 4 restrictions on census data.

The most important observations gathered through the performed test runs on random and US Census data are:

1. ADiT is up to 200 times faster than ARTO in case the search amount gets higher than the amount of peers in the network.
2. The system effort caused by ADiT is up to 8 times lower than the system effort caused by ARTO in case the search amount gets higher than the number of peers in the network.
3. The query response time of ARTO is in most cases worse than the query response time achieved with any of the presented ADiT heuristics.

Additionally, we found some characteristics appearing in almost all test runs:

– The enhanced heuristics is close to the approximated optimum gathered through the extensive tests on the *US Census Data (1990) Data Set*.
– The enhanced heuristics is better than all other presented heuristics, except in one single query.
– The enhanced heuristics is between 2 and 32 times faster than the heuristics always fetching *1 object* from each peer in parallel.
– The enhanced heuristics is about 3 to 8 times faster than heuristics fetching $\lceil \frac{k}{N} \rceil$ or $\lfloor \frac{k}{N} \rfloor$ objects from each peer in parallel.
– The enhanced heuristics is between 1.5 and 2.5 times faster than the heuristics fetching k *objects* from each peer in parallel.
– The basic heuristics and the heuristics fetching k *objects* from each peer in parallel turned out to be better than the other heuristics.

5   Conclusion

We discussed distributed top-k query processing from a new perspective. We motivated the need for an adaptive distributed top-k query processing approach (short ADiT) and defined two goal measures, namely (1) the system effort and (2) the query response time. Based on data gathered through extensive experiments we derived a heuristics which can be used to determine a separate fetch size for each peer. We compared the developed heuristics against other heuristics and against ARTO [21]. We could show that ADiT can accelerate the query response time and reduce the consumption of system resources significantly. Furthermore, we saw that the *enhanced heuristics* is in most cases close to the best system effort and query response time approximately determined upfront. Additionally, we found that a heuristics fetching more objects is usually the better choice since searching and transmitting a few more objects has much lower costs than sending an additional request. Last but not least the gains achieved with ADiT increase with the size of the peer to peer network and the number of requested results k.

References

1. UCI Machine Learning Repository, US Census Data 1990 (2012).http://archive.ics.uci.edu/ml/datasets/US+Census+Data+(1990)

2. Akbarinia, R., Pacitti, E., Valduriez, P.: Reducing network traffic in unstructured P2P systems using Top-k queries. Dist. Parallel Databases **19**, 67–86 (2006)
3. Akbarinia, R., Pacitti, E., Valduriez, P.: Best position algorithms for top-k queries. In: Proceedings of VLDB 2007 (2007)
4. Balke, W.-T., Nejdl, W., Siberski, W., Thaden, U.: Progressive distributed top-k retrieval in Peer-to-Peer networks. In: Proceedings of ICDE 2005. IEEE CS (2005)
5. Bruno, N., Chaudhuri, S., Gravano, L.: Top-k selection queries over relational databases: mapping strategies and performance evaluation. ACM Trans. Database Syst. **27**, 153–187 (2002)
6. Conner, W., Hwang, S., Nahrstedt, K.: Unified Framework for Top-k Query Processing in Peer-to-Peer Networks (2007)
7. Dabringer, C., Eder, J.: Adaptive Distributed Top-k Query Processing. CoRR 1606.01742 (2016)
8. Dabringer, C., Eder, J.: Efficient top-k retrieval for user preference queries. In: Proceedings of ACM Symposium on Applied Computing, SAC 2011. ACM (2011)
9. Dabringer, C., Eder, J.: Fast top-k query answering. In: Hameurlain, A., Liddle, S.W., Schewe, K.-D., Zhou, X. (eds.) DEXA 2011, Part II. LNCS, vol. 6861, pp. 144–153. Springer, Heidelberg (2011)
10. Eder, J., Dabringer, C., Schicho, M., Stark, K.: Information systems for federated biobanks. In: Hameurlain, A., Küng, J., Wagner, R. (eds.) Trans. on Large-Scale Data- and Knowl.-Cent. Syst. I. LNCS, vol. 5740, pp. 156–190. Springer, Heidelberg (2009)
11. Fang, Q., Yang, G.: Efficient top-k query processing algorithms in highly distributed environments. J. Comput. **9**(9), 2000–2006 (2014)
12. Fang, Q., Zhao, Y., et al.: Best position algorithms for top-k query processing in highly distributed environments. In: ICNDC 2010. IEEE (2010)
13. Frank, A., Asuncion, A.: UCI Machine Learning Repository (2010)
14. Frank, H., Eder, J.: Towards an automatic integration of statecharts. In: Akoka, J., Bouzeghoub, M., Comyn-Wattiau, I., Métais, E. (eds.) ER 1999. LNCS, vol. 1728, pp. 430–445. Springer, Heidelberg (1999)
15. Hagihara, R., Shinohara, M., Hara, T., Nishio, S.: A message processing method for top-k query for traffic reduction in ad hoc networks. In: Proceedings of International Conference on Mobile Data Management, MDM 2009. IEEE Computer Society (2009)
16. Hristidis, V., Papakonstantinou, Y.: Algorithms and applications for answering ranked queries using ranked views. VLDB J. **13**(1), 49–70 (2004)
17. Hua, M., Pei, J., Fu, A.W.C., Lin, X., Leung, H.-F.: Efficiently answering top-k typicality queries on large databases. In: Proceedings of VLDB (2007)
18. Ilyas, I.F., Beskales, G., Soliman, M.A.: A survey of top-k query processing techniques in relational database systems. ACM Comput. Surv. **40**(4), 1–58 (2008)
19. Mamoulis, N., Yiu, M.L., Cheng, K.H., Cheung, D.W.: Efficient top-k aggregation of ranked inputs. ACM Trans. Database Syst. **32**(3), 19 (2007)
20. Re, C., Dalvi, N., Suciu, D.: Efficient top-k query evaluation on probabilistic data. Data Engineering (2007)
21. Ryeng, N.H., Vlachou, A., Doulkeridis, C., Nørvåg, K.: Efficient distributed top-k query processing with caching. In: Kim, M.H., Unland, R., Yu, J.X. (eds.) DASFAA 2011, Part II. LNCS, vol. 6588, pp. 280–295. Springer, Heidelberg (2011)
22. Vlachou, A., Doulkeridis, C., Nørvåg, K., Vazirgiannis, M.: On efficient top-k query processing in highly distributed environments. In: Proceedings of SIGMOD (2008)
23. Zhang, Z., Hwang, S., et al.: Boolean + ranking: querying a database by k-constrained optimization. In: Proceedings of ACM SIGMOD (2006)

Basis Functions as Pivots
in Space of Users Preferences

Michal Kopecky, Marta Vomlelova, and Peter Vojtas[✉]

Faculty of Mathematics and Physics, Charles University in Prague,
Malostranske namesti 25, Prague, Czech Republic
{kopecky,vojtas}@ksi.mff.cuni.cz,
marta@ktiml.mff.cuni.cz

Abstract. Our starting motivation is a user visiting an e-shop. E-shops usually offer conjunction of sharp filter conditions and one attribute ordering of results. We use a top-k query system where results are ordered by a multi-criterial monotone combination of soft filter conditions.

For prediction of users' behavior, we introduce a class of basis functions with positive <u>L</u>inear combination of <u>T</u>riangular (soft) filters (LT). We prove that LT gives a unique representation of preferences. From database point of view LT act as a source for choosing pivots. From business perspective LT reflect aggregation of users' (soft) ideal values (choice points).

Our experiments use artificial data and are organized along variants of user's search habits, learning algorithms and evaluation measures. We argue that LT recommendations behave better with respect to order sensitive measures. This gives raise to a problem of pivot based indexing with order sensitive metrics.

Keywords: Business intelligence and analytics · Recommender systems · User preference learning · Pivot based indexing · Experiments · Evaluation measures

1 Introduction and Motivation

Recommendation systems have become a ubiquitous part of our web experience. They for example point us to interesting items to shop on e-commerce sites, they filter our news feeds on the social web, and they help us discover new music or movies on media streaming platforms. Correspondingly we could observe a huge interest in the computer science research community in this topic over the last decade.

Our starting motivation is a user visiting an e-shop. These usually offer conjunction of sharp filters and one attribute ordering of results. R. Fagin, A. Lotem and M. Naor in their JCSS 2003 paper [2] described a (middleware) top-k query system where each object in a database has m scores, one for each of m attributes that represent degrees of fulfillment of soft filters. To each object is then assigned an overall score that is obtained by combining the attribute scores using a fixed monotone combining rule. This approach enables multi-criterial ordering. We develop further this idea. Based on user's behavior we try to learn these filters and combination function.

© Springer International Publishing Switzerland 2016
M. Ivanović et al. (Eds.): ADBIS 2016, CCIS 637, pp. 45–53, 2016.
DOI: 10.1007/978-3-319-44066-8_5

Our contributions can be described as follows:

- For generalization (estimation) of users' behavior, we introduce a class of *basis functions* which reflect in a soft way users' ideal values (choice points).
- We study approximation properties of this class with respect to high dimensional space of all possible preferences and several (business motivated) metrics. We compare two methods for user preference learning. First one is based on using data pivots selected from basis functions, second by a new data mining.
- Our experiments show that the quality of approximation depends e.g. on number of objects user has visited, on weights of attributes, dimension of data space.

This work was motivated by paper of R. Fagin and his colleagues [2]. Some of our heuristics with attribute preference learning were motivated by [1]. Initial work on considering recommender systems evaluation separately for each user appeared already in [5]. Paper [3] shows that implicit preferences based on user behavior can be also translated to (fictitious) rating and hence dealing with explicit rating is sufficient.

We hope these will contribute to both database and information systems in general. Special points of interest can be found here for business intelligence and analytics, recommender systems and pivot based indexing.

In Sect. 2 we define basis functions, followed by sections concerned on general problems of user behavior approximation. Section 8 describes our experiments.

2   Preference Model and Class of Basis Functions

We represent preferences on a set $V \subseteq O$ of objects by *rating (scoring) function* $r: V \rightarrow [0;1]$, which assigns to objects $o \in V$ overall preference score $r(o) \in [0;1]$. We say an object o_1 *is more preferred* than object o_2 if $r(o_1) > r(o_2)$.

If V = "visited objects", then the rating r is called *observed rating*. If $V = O$, then the rating is said to be *total*. This rating is called *explicit*, if it comes directly from a user; *fictitious*, if it is deduced from user's implicit behavior (see [3]).

In [2] assume, that each object o has assigned m-many attribute scores $x_i^o \in [0;1]$. Combination function $t: [0;1]^m \rightarrow [0;1]$ is assumed to fulfill: $t(0,...,0) = 0$, $t(1, ...,1) = 1$ and t preserves ordering, i.e. if $x_j \leq_{[0,1]} y_j$ for all $1 \leq j \leq m$ then $t(x_1, ..., x_j, ..., x_m) \leq_{[0,1]} t(y_1,...,y_j,...,y_m)$. Because of this inequality we call this function *monotone*. The *overall score* of object o is $t(o) = t\left(x_1^o, ..., x_j^o, ..., x_m^o\right)$.

This class of preferences we call the **oFLN**-*class of object-preferences* (acting on objects). Because of the middleware setting, authors in [2] do not discuss how do (the server computed) score depend on query (search) and attribute values.

Now we introduce **dFLN**-*class of data-preference* acting on data cube. Assume, we have data with schema $[O, A_1, ..., A_m]$, where each attribute A_j has domain D_j and the preference is depending only on attribute values, i.e. for all $1 \leq j \leq m$ we have a vector **f** of attribute scoring functions $f_j: D_j \rightarrow [0;1]$. The dFLN-model assigns $\boldsymbol{a} \in \Pi D_j$ the overall score

$$r_{f,t}(\boldsymbol{a}) = t\big(f_1(a_1), \ldots, f_j(a_j) \ldots, f_m(a_m)\big).$$

We get the corresponding oFLN model by setting $f_j^o(o) = f_j(o.A_j)$, the corresponding vector is $\mathbf{f}^o$. The overall object scoring function corresponding to $r_{f,t}$ is given by formula

$$r_{f,t}^o(o) = t\big(f_1^o(o), \ldots f_j^o(o), \ldots, f_m^o(o)\big).$$

3 LT - Linear Combinations of Triangular Attribute Preferences

We are interested in a special subclass of dFLN-preference models, which are generated from data (objects' attributes) over numeric domains (distinguished by shape of attribute preferences and properties of combination function).

Assume $D_j = [a_j; b_j]$. First component of our model aims to model soft filtering conditions. An attribute preference f_j is a triangular (trapezoidal/bell shape analogously) scoring function defined by an ideal (choice point) value $c_j \in [a_j; b_j]$, i.e.

$$f_j(x) = \min\big\{(x - a_j)/(c_j - a_j), (b_j - x)/(b_j - c_j)\big\}.$$

Combination function is weighted average $t_w: [0; 1]^m \rightarrow [0; 1]$ with non-negative coefficients $\mathbf{w} = (w_1, \ldots, w_m)$. Then r_{fw} is called an *LT-preference* (rating). Class of all such total ratings will be denoted as LT.

In our recommendation, we will look for user models from this class. We develop procedures, which for given user behavior find (best) user model from class LT.

4 Uniqueness of Continuous LT Preference Representation

One of desired properties of class of basis functions is uniqueness – different basis functions have to represent different generalizations. E. Huellermeier rose following question: Does LT, in aspiration to become a reasonable class of basis functions, fulfills uniqueness? Following result was first presented at [4].

Observation. Assume D_j, $\mathbf{f} = \langle f_j \rangle$, c_j are as before. Assume moreover, there is $r_{g,s}$ with choice points $d_j \in D_j$ and triangular local preferences $\mathbf{g} = \langle g_j \rangle$, $g_j: D_j \rightarrow [0,1]$.

Let $t, s: [0,1]^m \rightarrow [0,1]$ are two aggregation functions, such that they violate uniqueness, i.e. $\exists\, 1 \leq j_0 \leq m$ such that $c_{j0} \neq d_{j0}, \forall x = (x_1, \ldots, x_j, \ldots, x_m) \in \Pi D_j$

$$r(x_1, \ldots, x_j, \ldots, x_m) = t\big(f_1(x_1), \ldots, f_j(x_j), \ldots, f_m(x_m)\big)$$
$$= s\big(g_1(x_1), \ldots, g_j(x_j), \ldots, g_m(x_m)\big).$$

We claim, that then for all $(x_1, \ldots, x_{j0-1})$, $(x_{j0+1}, \ldots, x_m)$ we have the function

$\lambda x_{j0} r\left(x_1, \ldots, x_{j0-1}, x_{j0}, x_{j0+1}, \ldots, x_m\right)$ is constant on $[a_{j0}, b_{j0}]$, that is, the value of the function r does not depend on attribute A_{j0}.

Proof. Assume, that $j_0 = 1$, $a_1 < c_1 < d_1 < b_1$. Let $\mathbf{c} = (c_2, \ldots, c_m)$, $\mathbf{f} = (f_2, \ldots, f_m)$, $\mathbf{d} = (d_2, \ldots, d_m)$, $\mathbf{g} = (g_2, \ldots, g_m)$ and $\mathbf{x} = (x_2, \ldots, x_m)$ is arbitrary.

We will iteratively use monotonicity of triangular shape of local preferences and the fact that combination functions preserve ordering. To start, notice (see Fig. 1.) that $\forall x \in [a_1; b_1]$ $r(x,\mathbf{x}) \leq r(c_1, \mathbf{x})$ and $r(x, \mathbf{x}) \leq r(d_1, \mathbf{x})$, hence $r(c_1,\mathbf{x}) = r(d_1,\mathbf{x}) = m_{\mathbf{x}}$ are global maximum of the function $r(., \mathbf{x})$. We will prove that $\forall x \in [a_1, b_1]$ $r(x,\mathbf{x}) = m_z$. As $c_1 < d_1$, $f_1(c_1) > g_1(c_1)$, there is an $y_1 > d_1$ such that $g_1(y_1) = g_1(c_1) < 1$, then for all $x \in [c_1; y_1]$, $g_1(x) \geq g_1(c_1)$, $r(c_1,\mathbf{x}) = s(g_1(c_1),\mathbf{g}(\mathbf{x})) \leq s(g_1(x), \mathbf{g}(\mathbf{x})) = r(x,\mathbf{x}) \leq r(c_1, \mathbf{x}) = m_{\mathbf{x}}$, hence $r(., \mathbf{x})|[c_1, y_1] = m_{\mathbf{x}}$, i.e. $r(., \mathbf{x})$ is on $[c_1, y_1]$ constantly equal to $m_{\mathbf{x}}$.

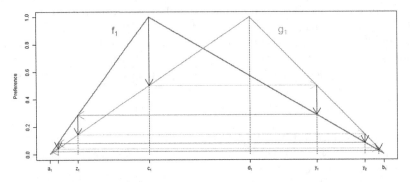

Fig. 1. Illustration for the proof of uniqueness of LT preference representation

Now we can lead our argument in the opposite direction with $z_1 < c_1$ such that $f_1(z_1) = f_1(y_1)$ and so on.

In case $a_1 = c_1$, then $g_1(c_1) = 0$ and $y_1 = b_1$ and the argument that $r(., \mathbf{x})$ is on $[c_1; y_1]$ constantly equal to $m_{\mathbf{x}}$ goes through. Note, we did not use linearity of aggregations (we need only continuity for end points).   □

5   Estimating General Customer's Preferences by LTs

Our recommendation is content based and learned ratings usually act on data cube ΠD_i. We present a multi-customer preference model.

Assume we have a set of users U and for each $u \in U$ the set $V^u \subseteq O$ of visited objects and corresponding observed rating $r^u: V^u \rightarrow [0; 1]$. We divide V^u to disjoint union $V^u_{train} \cup V^u_{test}$ of training and testing examples. This implies, we have also r^u_{train}: $V^u_{train} \rightarrow [0; 1]$ and $r^u_{test}: V^u_{test} \rightarrow [0; 1]$.

Content based learning assumes knowledge about actual data $d(O) \subseteq \Pi D_i$ (d maps objects to data cube). We transform training examples to data cube by $r^u_{train}(d^{-1})$. Our task is to find a total rating $r^u_{rec}: d(O) \subseteq \Pi D_i \rightarrow [0; 1]$ such that

– either comparing ratings acting on the data cube $\rho\left(r_{test}^u(d), r_{rec}^u\right), \tau\left(r_{test}^u(d), r_{rec}^u\right)$,

– or comparing ratings acting on objects $\rho\left(r_{test}^u, r_{rec}^u(d^{-1})\right), \tau\left(r_{test}^u, r_{rec}^u(d^{-1})\right)$ is best possible with respect to metric ρ (τ order sensitive resp.).

6   Pivots in the Space of All Ratings

The data space of all possible ratings/customers is the set of all possible mappings from O into [0; 1], denoted by $[0; 1]^O$ ($[0; 1]^{\Pi D_i}$ resp.). It is a high dimensional space, where the number of dimensions is either $|O|$ or $|\Pi D_i|$, resp.

Set of potential observed ratings can be described as the sets of ratings $[0,1]^V$ (mappings from V "small" subsets of O to [0,1]).

We are going to use the idea of pivots. Space LT of LT-ratings is very efficient representation of preferences. We need only vectors of choice points c_i and weights w_i. So our idea is to find pivots in $P \subseteq LT \subseteq [0; 1]^{\Pi D_i}$.

We leave for future work how to index ratings and find pivots with respect to actual data and various business related metrics, measures, evaluations like RMSE, nDCG, Kendal τ, 1-hit, Next-k etc. Here we work with idea of "universal" pivots constructed in $P \subseteq LT \subseteq [0,1]^{\Pi D_i}$.

7   Methods of Learning Generalizations Based on Observed Ratings

We present two approaches which were tested:

Method 1 pivot based – Estimate user behavior by the closest pivot.

– Fix a problem domain and a set of pivots $P \subseteq LT \subseteq [0; 1]^{\Pi D_i}$.

Given a user u and an observed user rating $r^u: V^u \to [0;1]$ in case of online experiments or $r_{train}^u: V_{train}^u \to [0; 1]$ in case of offline experiments,

– Find $p^u \in P$ where $\rho(r^u, r_{pu}), \tau(r^u, r_{pu})$ is best possible. Note, that pivot is in LT and it induces a global rating r_{pu}

For an unseen object $o \in [0; 1]^{\Pi D_i} \setminus V^u$ ($[0; 1]^{\Pi D_i} \setminus V_{train}^u$ resp.)

– Recommend $r_{pu}(o)$, eventually recommend top-k wrt $r_{pu}(o)$
– Check the quality of recommendation either on future behavior in case of online experiments or on testing examples in case of offline experiments wrt ρ, τ resp. This method will serve as a baseline for our further investigations.

Method 2 model based – Two step data mining.

– Fix a problem domain and quality measures ρ, τ resp.

Given an observed user rating $r^u: V^u \to [0; 1]$

– Step 1. Find c_j, $\mathbf{f} = <f_j>$, $f_j \in LT(D_i)$ attribute preferences

- Step 2. Find **w** weights for t^w linear combination function

  For an unseen object $o \in [0; 1]^{\Pi D_i} \backslash V^u$ ($[0; 1]^{\Pi D_i} \backslash V^u_{train}$ resp.)
- Recommend $r_{f,w}(o)$, eventually recommend top-k wrt $r_{f,w}(o)$
- Check the quality of recommendation either on future behavior in case of online experiments or on testing examples in case of offline experiments wrt. ρ, τ resp.

Step 1: For each dimension D_i we selected $\hat{c}_{ij}$ the best estimate of the choice point based on training data by minimizing $RSS = (y_j - f_i(x_{ij}))^2$, see Fig. 2.

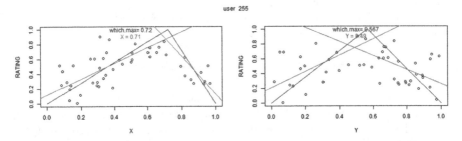

Fig. 2. One of RMSE oriented heuristics for learning attribute preferences (see also [1] and step 1 of our data mining method). Note that X, Y have normalized domains.

Step 2: We computed weight estimates by linear regression without intercept of input features f_i from step 1 on output variable RATING.

8   Data – Experiments – Metrics – Results

To check whether these ideas are viable, we provide a series of experiments on artificial data. We start with a seemingly trivial problem, our user's behavior is described by a LT model and we try to find the closest pivot (which is also from LT). The problem is that we know about the user only her/his evaluation of a small part of data. For our test we used artificial randomly generated data of items, users and ratings with several alternatives.

Items. For our test we used artificially randomly generated 3D data in a numeric domain (attributes denoted x, y, z), 10.000 data points with normal distribution with centers in different parts of data cube, forming several clusters.

Users = Rating–shape and Frequency. Generating users we concentrated on user's frequency with search habits and shape of user's preference.

Shape. Triangular shaped: users with id's in range 1–300 have an LT preference with randomly generated choice points (ix, iy, iz) and weights (w_x, w_y, $w_z = 1 - w_x - w_y$).

Bell shaped: users (id's 1000–3125) rating by normal distribution with axes not parallel with attribute axes and instead of weights there is a variation matrix.

Frequency and Search Habits. Training examples can be considered as those coming from the first search phase in which users explicitly rate. Test examples can be considered as those coming from the second search phase where we recognize two basic search habits – *random* and *targeted*. Random users visit and rate items also in the second phase randomly, while targeted user behaves in the second phase by his/her preferences. During generation of users we concentrated also on user's diligence modeled by probability p_{train} that a data point falls into training and p_{test} into testing examples. We tested three types of users with probabilities in following intervals

Diligent random users, with $p_1 = p_{train} \in [0.01; 0.05]$, $p_2 = p_{test} \in [0.01; 0.05]$, id's 1–100 and 1000–1125 (#visited items 100–500).

"Bewildered" random users, $p_1 = p_{train} \in [0.001; 0.005]$, $p_2 = p_{test} \in [0.01; 0.05]$, id's 101–200 and 2000–2125, (visited/rated 10–50 training and 100–500 test items).

Lazy random users, $p_1 = p_{train} \in [0.001; 0.005]$, $p_2 = p_{test} \in [0.001; 0.005]$, id's 201–300 and 3000–3125 visited/rated 10–50 chosen items in both training and testing phases.

We will note users as *t-diligent*, *t-bewildered* and *t-lazy* for triangular-shaped users and *b-diligent*, *b-bewildered* and *b-lazy* for bell-shaped ones.

Metrics. Basic challenge we want to tackle here is comparison of classical RMSE metrics with order sensitive metrics in the area of user preferences.

RMSE. We calculate classical root of average of square error between correct and computed user rating separately for each user. See results in Fig. 3.

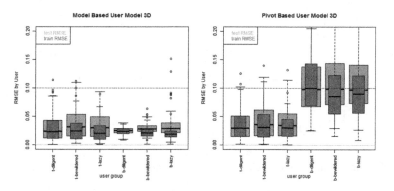

Fig. 3. RMSE for random users – model-based vs. pivot based 3D methods compared.

Order Sensitive Metrics. Assume we have two ordering $<_1$ is generated by observed user rating and $<_2$ is generated by our estimation of the same set and $k > 0$. Then the number of top-k agreements between $<_1$ and $<_2$ is defined as the size of intersection of top-k($<_1$) and top-k($<_2$), in formula #top(k, $<_1$, $<_2$) = $|$ top-k($<_1$) $\cap$ top-k($<_2$)$|$, see Fig. 4. Surprisingly pivot based methods shows better results than data mining method. Green line denotes the direction of measure being better.

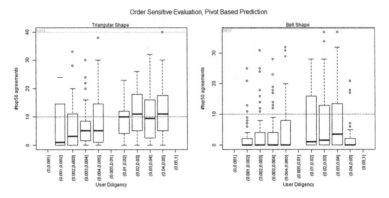

Fig. 4. Measuring size of intersection of top-50.

Second measure is motivated by an idea, that a user would be grateful even in a case when at least one recommended object fulfills his/her preferences, 1st-hit $(<_1, <_2) = \min \{k: \text{top-}k(<_1) \cap \text{top-}k(<_2) \neq \varnothing\}$, see Fig. 5. It shows only targeted users, triangular shaped and diligent users perform better than bell shaped and lazy ones.

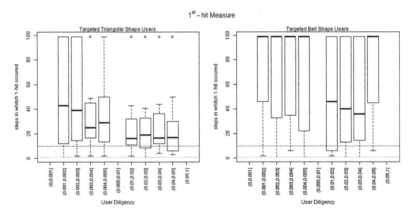

Fig. 5. The minimal size of predicted top-k set to contain at least 1 hit from real top-k.

9   Conclusions, Future Work

To conclude, we state that our aim to study the user preference learning problem in a general setting – that is investigating the whole space of all possible observed rankings was presented in a form of experiments of different methods on randomly generated items, users and ratings data. By our opinion, results show that this approach is worth of further research.

In general we can conclude that triangular-shaped users are easier to predict than bell-shaped. Similarly, diligent users are easier to predict than lazy ones. Behavior estimation quality depends on metric, random users are easier for RMSE and targeted users for order sensitive metrics.

Please note, that as a side product of our approach, the use of pivots for collaborative filtering can contribute to the cold start problem.

Acknowledgements. We announce partial support of Czech grants P103-15-19877S, 16-09103S and P46.

References

1. Eckhardt, A.: Similarity of users' (content-based) preference models for Collaborative filtering in few ratings scenario. Expert Syst. Appl. **39**(14), 11511–11516 (2012)
2. Fagin, R., Lotem, A., Naor, M.: Optimal aggregation algorithms for middleware. JCSS **66**(4), 614–656 (2003). (Preliminary version ACM PODS 2001)
3. Peska, L., Vojtas, P.: Using implicit preference relations to improve recommender systems. J. Data Semant., 1–16 (2016). http://dx.doi.org/10.1007/s13740-016-0061-8
4. Vojtas, P.: Linear combinations of triangular attribute preferences provide unique representation of user preferences. Seminar of Prof. E. Huellermeier, Marburg, 26 February 2013
5. Vojtas, P., Kopecky, M., Vomlelova, M.: Understanding transparent and complicated users as instances of preference learning for recommender systems. In: Kofron, J., Vojnar, T. (eds.) MEMICS 2015. LNCS, vol. 9548, pp. 23–34. Springer, Heidelberg (2016). doi:10.1007/978-3-319-29817-7_3

JBD Generator: Towards Semi-Structured JSON Big Data

Roman Betík and Irena Holubová[✉]

Department of Software Engineering,
Charles University, Prague, Czech Republic
rbetik@gmail.com, holubova@ksi.mff.cuni.cz

Abstract. This paper describes a tool for generation of synthetic semi-structured JSON Big Data, called *JBD generator*. Its main focus is on parallel execution of the generation process while preserving the ability to control the contents of the generated documents. It can also accept samples of real-world data characterizing the target synthetic data and is also capable of automatic creation of references between JSON documents. The results of experiments with the data generator exploited for the purpose of testing database *MongoDB* describe its added value.

Keywords: Big Data · Data generator · MongoDB · JSON documents

1 Introduction

Nowadays, computer systems of many companies work with very large data sets called *Big Data*. This situation has introduced many new problems how this data should be stored, processed, searched, transferred, visualized etc. With regards to the features of the data, so-called $NoSQL^1$ databases have been introduced. One of the most complex data structures are offered by NoSQL *document databases* which store data in the form of semi-structured or structured documents, in formats like XML [2] or JSON [1]. Since there naturally exist several competing implementations, the developers must choose the most suitable one for their solution.

Comparing and testing such systems involve usage of test data, either real-world data from a similar application or synthetically generated data. Obtaining real-world data is often difficult due to its size and/or privacy concerns. Therefore, we have analyzed suitable XML data/Big Data/JSON data generators and described their (dis)advantages [4]. Our consequent main focus is generation of synthetic Big Data in the JSON format, the most popular format in this area. Then we demonstrate the abilities of the proposed *JBD (JSON Big Data) generator* using $MongoDB^2$, the most popular representant of document stores.

Supported by the grant SVV-2016-260331.

¹ http://nosql-database.org/.
² https://www.mongodb.org.

© Springer International Publishing Switzerland 2016
M. Ivanović et al. (Eds.): ADBIS 2016, CCIS 637, pp. 54–62, 2016.
DOI: 10.1007/978-3-319-44066-8_6

The rest of the text is structured as follows: Sect. 2 describes the architecture of the proposed generator and Sect. 3 provides the results of experiments depicting its features. Section 4 concludes and outlines possible future work.

2   JBD (JSON Big Data) Generator

The top-level components of the JBD are shown in Fig. 1. `JsonGenerator.`
`Master.Window` (`Master` from now on) has the main role in JBD. Here the data generation process starts and the properties of the data generation process are configured. It also shows the progress. `JsonGenerator.Master.Generator.`
`Client` (`Client` from now on) runs on every server responsible for data generation. The role of this application is to receive tasks from `Master`, delegate it to the actual data generator and then notify `Master` about the results. The actual data generator and schema analyzer component `node.js` is also present in multiple instances and has one-to-one relationship with `Client`. It infers the schema from the input documents and performs the data generation.

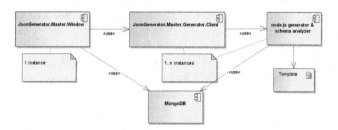

Fig. 1. Main components of JBD

MongoDB is used as a storage for both input and generated documents. This allows us to implement the references between documents without a special communication protocol. Testing another database requires to abstract the interface for the database and perform the lookup for references separately. And the same strategy can also be applied to the document saving stage.

2.1   Templates

JBD uses the same templates for data generation as used by *MongoDB-Datasets*[3], an open-source *Node.js*[4] application designed to generate data. The template is a JSON document with parts written in JavaScript in double curly braces {{}} (see example below). We added additional meta data for JBD so that it knows where to store the generated data when it generates data using multiple templates. The following example shows this meta data written as the value of the field _$db.

---

[3] https://github.com/mongodb-js/datasets.
[4] https://nodejs.org/.

```
"_$db": {
    "sourceDb": "local",
    "sourceCollection": "test",
    "targetDb": "test",
    "targetCollection": "people"
},
"firstName": "{{ chance.first() }}",
// ...the rest of the template...
```

These settings instruct the data generator where to save the results of the data generation. This structure can be extended if the data generator needed additional data during the data generation phase.

2.2 Schema Inferring

Writing a template for data generator can be sometimes a tedious work. Thus our solution uses a *Node.js* application *mongodb-schema*[5]. It produces a probabilistic schema for a document collection in *MongoDB*.

The inferred information includes the names and types of all fields and also the probability of a field being included in the document. There are also all possible values stored that were present in the input documents. The classes correspond to parts of the JSON document. But, unfortunately, JSON has only basic types like `number`, `boolean` and `string`, whereas our documents may involve more specific data types. Therefore we use functions from libraries *Chance*[6] and *faker.js*[7] which can produce various random data types, such as names of people, words, sentences, dates like birthdays, URLs etc.

2.3 Data Generation

The data generator is implemented in JavaScript using *Node.js* framework. From the given template it constructs an in-memory representation of the documents, i.e., the tree structure of JavaScript objects which also contains the part of the template written inside the double curly braces {{}}. These parts of the template are compiled into evaluable JavaScript functions using library *underscore.js*[8]. It is therefore possible to inject any kind of JavaScript code into the template if the data generator references the functions.

The following list summarizes the capabilities of the data generator:

- The data generator can generate basic JSON data types.
- The data generator can generate random data using third party solutions.
- The data generator provides type conversions specifically for *MongoDB* (`Date`, `Number`, `ObjectID`, `Timestamp`).
- It is possible to use `this` keyword in template expressions to reference other generated values.

---

[5] https://github.com/mongodb-js/mongodb-schema.
[6] http://chancejs.com/.
[7] https://github.com/FotoVerite/Faker.js.
[8] http://underscorejs.org/.

– It is possible to hide or show fields based on a JavaScript expression.
– It provides the following methods:
  • _$size – the total number of generated documents in the current run
  • _$index – the index of the current document, starting from 0
  • counter([id], [start], [step]) – the underlying counts are accessible anywhere in the outmost document so that it is possible to use the same counter consistently regardless its position
    * id – the index of the counter to use, default is 0
    * start – the first count, default is 0
    * step – increment of each count, default is 1
  • util.sample(list, [n]) – chooses n items from the supplied list of values (used in arrays)

The current implementation generates documents either directly to *MongoDB* or it can save them to files on disk.

Document References. *MongoDB* offers two methods for referencing: (1) *manual references*, where the application saves the _id field of one document in another document as a reference, and (2) *DBRefs*, i.e., references from one document to another using the value of the _id field of the referenced document, collection name, and, optionally, its database name. *MongoDB* website suggests the use of manual references. Therefore, these references are also supported in JBD.

MongoDB-Datasets does not support the generation of the references between documents so we had to add this functionality to the node.js application. The data generator then chooses a random document from the referenced collection and writes its ObjectId to the referencing document. In order to support the random document selection, the target collection must have a special field rnd defined in its template.

The method we described so far will generate references to the whole collection. If we want to limit the set of documents which are referenced, it is possible to alter the generation of the rnd field so that it is hidden based on an arbitrary condition. Then the data generator chooses only from this restricted set of documents. An example can be in the case when the rnd field is hidden for people with the degree 'Bc.' so candidates for references will not include these people:

```
"rnd" : "{{NumberLong(chance.natural())}}"
// ... the rest of the template ... or optional rnd field:
"rnd": "{{NumberLong(chance.natural())}}{{ hide(this._$degree =='Bc.') }}",
"_$degree": "{{ util.sample(['Bc.','MSc.','Ph.D.','']) }}"
```

JBD currently supports generation of references only when it is used with *MongoDB*. Output to files requires that the template contains no references. JBD also does not support automatic inferring of the references between documents when analyzing existing data sets. There is no standard way how references are implemented in JSON documents so only a heuristic method can be used. We suggest the following method: First, the user specifies which collections (s)he wants to use for the analysis. The algorithm analyzes each collection and marks every field with `ObjectId` type as a possible candidate for a document reference. Then for each document in the collection and each candidate field it tries to get the document with the corresponding `_id` from all the other collections. If all of the `ObjectId`s (or a majority) would point to the same collection, the algorithm would assume this as a reference to that particular collection. The user can then verify the result and correct it manually, if necessary.

2.4 Task Delegation and Parallelism

The important part of our work was to enable parallel data generation. We wanted to have independent data generation units which could run in many instances. These have to be somehow managed and hence our data generator would have one main control part – a *coordinator*. For cross-platform interprocess communication we used the *Eneter Messaging Framework*[9] which offers several message passing mechanisms and is easy-to-use. We used its broker service to create a publish-subscribe model where clients subscribe to the server and notify it that they are ready to receive tasks. The server then sends messages to all the clients at once for simplicity. The task delegation is solved at the application layer. The whole process shown in a simplified form in Fig. 2.

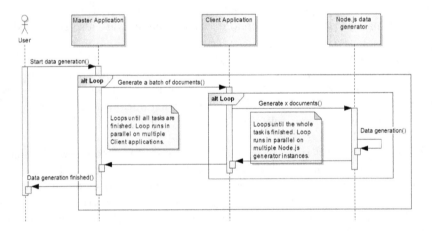

Fig. 2. Diagram of task delegation in JBD

3   Experiments

We created 3 different templates, corresponding to collections, which model a database of people, their phones and messages (`people`, `phones`, `messages`) sent between them, and 1 template to generate a collection of documents that model a blogging platform (`cms`, `logs`). Then we defined 4 different test configurations which alter the creation of collections, sharding settings, and indices:

---

[9] http://www.eneter.net.

- *Test 1*: All collections are created without indices except the `rnd` field in `people` collection for random document selection. Sharding is not enabled.
- *Test 2*: Contrary to *Test 1* sharding is enabled using the `_id` field.
- *Test 3*: In addition we included hashed indices on `_id` fields. Sharding is enabled using these fields with the option `hashed` giving random distribution of data across the shards.
- *Test 4*: In addition there are several indices created for various fields to suit some of the test queries. Sharding is enabled on hashed `_id` fields.

Each of these tests is performed using 1, 2 and 3 worker generators and for each number of generators we generated 1 000 000 documents for every collection.

3.1   Test Queries

For every test we executed the testing queries three times and measured the execution time. The queries[10] involve testing of filtering (QP1), optional values (QP2), working with arrays (QP3), value ranges (QP4), fields inside object fields (QP5), `ObjectId` fields (QM1-2), intervals (QPh1), filtering on array fields (QC1), sorting (QC2), and date range (QL1).

3.2   Testing Environment

The testing environment consisted of 12 virtual servers – 3 x application server (App1-3), 3 x *MongoDB* configuration server, 2 x *MongoDB* query router, 4 x *MongoDB* database server. The deployment was based on the architecture proposed by the authors of *MongoDB* [3]. Every worker created 12 independent threads and each of these threads was generating data and writing it directly to the database in batches of 1 000 documents (as imposed by *MongoDB*).

Servers App1–3 hosted `Client` together with the `node.js` data generator and schema analyzer. We controlled and monitored the data generation process from the App1 server where we executed `Master`. The rest of the servers were all part of the *MongoDB* cluster and we did not modify it. We used *MongoDB* version 3.0.4. All insert operations used acknowledged write concern.

3.3   Generation of 1 000 000 Documents

Figure 3 summarizes the generation of 1 000 000 documents using 1 client generator (i.e., 12 threads). Figure 4 shows the same results for 3 client generators (i.e., 36 threads). It is obvious that adding another client generator increased the data generation rate. The biggest difference was between using 1 and 2 client generators (the data generation was approximately twice as fast). Addition of another client data generator sometimes even made it worse as it probably overloaded the database. Sharding and indices influenced the data generation rate slightly, whereas it slowed down the data generation in majority of our tests.

---

[10] For the full list of queries and their listings see [4].

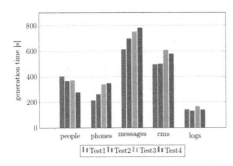

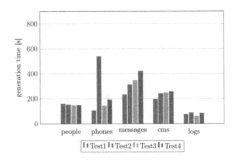

Fig. 3. 1 client generator   **Fig. 4.** 3 client generators

Data Generation Rate. Table 1 contains the computed generation rates for each test we performed while generating 1 000 000 documents. The maximum rate at which the data generator was able to produce data was around 10MB/s (1 generator, 12 threads). When the generation process included also references, this rate dropped significantly – in some cases even under 1MB/s.

Additional generator (additional 12 threads) raised the maximum capacity to almost 20MB/s. It also improved the generation with references – it roughly doubled this rate. The third data generator did not improve the overall data

Table 1. Data generation rates for 1 million documents

Collection	Generators	Test 1 [kB/s]	Test 2 [kB/s]	Test 3 [kB/s]	Test 4 [kB/s]
people	1	7 909.96	8 675.53	8 523.13	11 442.72
phones	1	1 082.66	889.30	687.86	671.90
messages	1	1 266.35	1 114.56	310.93	993.93
cms	1	2 509.14	2 485.38	2 034.24	2 137.06
logs	1	3 391.48	3 658.40	2 924.32	3 447.86
people	2	20 593.85	15 706.42	15 096.75	17 707.92
phones	2	2 230.02	1 508.87	1 323.54	1 271.07
messages	2	2 997.97	2 149.78	2 009.67	1 909.62
cms	2	6 122.13	4 542.10	4 504.30	3 792.91
logs	2	6 734.01	6 276.57	6 445.93	6 935.54
people	3	19 700.85	20 447.12	21 470.02	21 005.17
phones	3	2 186.95	433.42	1 604.74	1 206.18
messages	3	3 328.29	2 476.22	2 223.03	1 841.14
cms	3	6 282.85	5 082.09	4 981.60	4 814.65
logs	3	6 270.25	5 342.91	7 959.14	5 678.19

generation speed. In some cases, it was even worse. We also monitored the load on the database server – 2 generators were enough to fully load it.

Query Performance. We also measured the execution of the sample queries. Summarization of the measured values is present in Figs. 5, 6, 7 and 8. If we compare the query performance between *Test 1* and *Test 2*, we can see that the sharding helped to speed up the query execution time. It was roughly twice as fast. Query performance measured after *Test 3* also showed an overall improvement. An interesting finding, however, is that queries we executed after *Test 4* were slower in cases when there were no indices created for the queries. Query execution times for fields which were indexed were much faster.

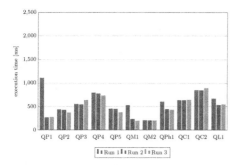

Fig. 5. 1 000 000 documents, Test 1

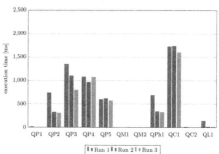

Fig. 6. 1 000 000 documents, Test 2

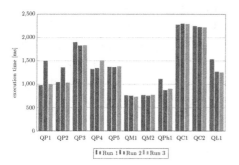

Fig. 7. 1 000 000 documents, Test 3

Fig. 8. 1 000 000 documents, Test 4

4   Conclusion

The proposed generator is able (1) to run in parallel on multiple servers producing different volumes of data at different rates, (2) to infer a schema of an existing set of documents, and (3) to create references between the generated documents. To the best of our knowledge there are no similar solutions that could be used with the same small amount of effort. However, JBD is a prototype which focuses on the core problems (i.e., data generation, references, parallelism), less on the overall usability in terms of the user interface, support for other document databases and formats.

References

1. ECMA-404 The JSON Data Interchange Standard (2015). http://json.org/
2. Extensible Markup Language (XML) 1.0 (5th edn.). W3C (2013)
3. Production Cluster Architecture. MongoDB Inc (2015). http://docs.mongodb.org/manual/core/sharded-cluster-architectures-production/
4. Betik, R.: Automatic Generation of Synthetic XML Documents, Master Thesis, Charles University in Prague (2015). http://www.ksi.mff.cuni.cz/~holubova/dp/Betik.pdf

Skyline Algorithms on Streams of Multidimensional Data

Alexander Tzanakas, Eleftherios Tiakas[✉], and Yannis Manolopoulos

Department of Informatics, Aristotle University, 54124 Thessaloniki, Greece
{tzanakas,tiakas,manolopo}@csd.auth.gr

Abstract. We compare three algorithms for skyline processing on streams of multidimensional data with centralized processing, namely, the *LookOut*, *Lazy* and *Eager* methods, with different dataset types and dimensionalities, data cardinalities and sliding window sizes. Experimental results for time performance and memory consumption are presented. In addition, the problem of computing the exclusive dominance region in higher dimensions is reviewed and a novel correct solution is proposed.

1 Introduction

Skyline queries stem in applications where user preferences determine the result. More formally, if a dominance realtionship in a dataset is defined, a skyline query returns the objects that cannot be dominated by any other object. In other words, if the dataset contains multidimensional objects, an object dominates another one if it is as good in all dimensions, and better in at least one dimension.

Skyline computation algorithms are divided into two categories. The first category consists of algorithms that inspect static data; there are no insertions or deletions while executing the algorithm. For example, a user wants to pick a hotel based on the price and its distance from the beach. The user defines in the dominance relationship that the lower price and the smaller distance, the better. In Fig. 1a, the X-axis depicts the distance from the beach, the Y-axis depicts the price, whereas the zigzag line represents the skyline. But hotel rooms are booked by other users and become unavailable, so a mechanism for removing the unavailable rooms, or inserting new ones is needed. This case of skyline computation is called *continuous*, because the skyline is continuously calculated and updated. Figures 1b–c depict the change of skyline after deleting object x.

Skyline queries have been examined thoroughly in the past. Börzsönyi et al. proposed the use of the skyline operator [1]. Tan et al. used bitmaps and B$^+$-trees to compute the skyline [11]. Kossmann et al. developed an algorithm that enables users to include their preferences at execution time [4]. Chomicki et al. proposed the *SFS* algorithm that uses a monotone function to compute the skyline mainly in relational data [2]. Papadias et al. computed the skyline using the distance from the axis origin with the use of spatial indexing techniques [9]. Skyline algorithms on data streams assuming various environments have lastly

© Springer International Publishing Switzerland 2016
M. Ivanović et al. (Eds.): ADBIS 2016, CCIS 637, pp. 63–71, 2016.
DOI: 10.1007/978-3-319-44066-8_7

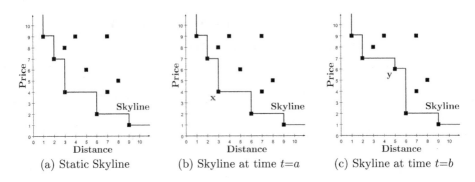

(a) Static Skyline   (b) Skyline at time $t{=}a$ (c) Skyline at time $t{=}b$

Fig. 1. Skyline examples

received increased attention [5]. For example, algorithms for skyline queries over data streams and centralized processing are: the *LookOut* algorithm [8], *Lazy* and *Eager* algorithms [12]. Also, recent algorithms for skyline queries over data streams and distributed processing are: the *SWSMA* algorithm [15] and the *Two-phase solution* [6].

Even though a lot of work has been done for assorted instances of skyline queries, not much has been done for a rather global and exhaustive evaluation of the skyline algorithms in centralized environments. This paper tries to address these issues by comparing the algorithms for skyline queries over data streams and centralized processing that are widely used: the *LookOut*, *Lazy*, and *Eager* algorithms. In particular, the contribution of this paper lies in:

1. thoroughly evaluating the above skyline algorithms over data streams under several multidimensional datasets, dataset cardinalities and sliding window (*SW*) sizes
2. explaining the deficiency of the *Lazy* algorithm during the computation of the exclusive dominance region in high dimensions (see Sect. 2.1), and proposing a solution. This results in an improvement, which makes the algorithms to work more properly and remain efficient for high dimensionalities
3. establishing a simple solution that can be applied in any skyline algorithm over data streams, which uses the exclusive dominance region.

Further insights can be found in the full version of this paper [13].

2   Continuous Skyline Computation Algorithms

Here, we examine the *LookOut* [8], *Lazy* and *Eager* methods [12]. First, certain implementation aspects are examined and then they are evaluated on the basis of execution time, memory allocation and *SW* size.

2.1   The Lazy Algorithm

The *Lazy* algorithm has been presented by Tao and Papadias in [12]. Changes in skylines may arise if: (i) a new tuple is inserted in the database, or/and (ii) an object expires and has to be removed from the database. The *expiration time* of an object equals *arrival time + sliding window size*. The *Lazy* algorithm uses a pre-processing module (*L-PM*) and a maintenance module (*L-MM*). When a tuple r is inserted, the *L-PM* module checks if it is dominated by a tuple of the current skyline. E.g. in Fig. 2a the arrival of tuple f at time $t=14$ does not affect the current skyline, because f is dominated by d. Thus, f is saved with the objects not currently used in the skyline, but possibly to be included at a later time. The database which stores the *inactive* data is called *DBrest*, whereas the database which stores the skyline is called *DBsky*.

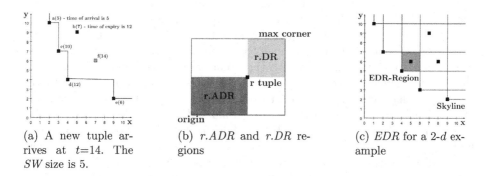

(a) A new tuple arrives at $t=14$. The *SW* size is 5.

(b) $r.ADR$ and $r.DR$ regions

(c) *EDR* for a 2-d example

Fig. 2. Skyline algorithms' features

If the incoming object dominates some of the skyline objects, it is stored in *DBsky*, whereas the dominated data are deleted as they will never appear again. The algorithm also defines two regions of a tuple r: (i) the dominance region $r.DR$ with starting point the coordinates of the object r and ending point the maximum coordinates that can appear, and (ii) the anti-dominance region $r.ADR$, which covers a region that spans from the start of the axes to the object itself. Figure 2b depicts the shape of $r.DR$ and $r.ADR$ in a 2-d setting.

When a tuple arrives, we check if any of the skyline objects are in the $r.ADR$ region. In contrast, the *Lazy* algorithm performs an $r.DR$ query to find the objects belonging in the dominance region of the new tuple. If an object is found in the $r.ADR$ region, then the new tuple is stored in the *DBrest* database, where it will stay until it is included in the skyline or it expires. On the other hand, if there are objects in the $r.DR$ region, they are expunged from the system and the new one is inserted in the skyline. The expiration time for the skyline is set to the lowest value found in it.

The *L-MM* module maintains the data existing in the database already. For this reason, it is executed at the time specified by the *L-PM* module, i.e. when an object expires and has to be deleted from the skyline. The algorithm removes the specific object and removes the objects that are stored in *DBrest* and have expired already. Then the skyline is recomputed, only for the objects that are dominated exclusively by a tuple r, which is about to be deleted. In Fig. 2c the

Exclusive Dominance Region (*EDR*) for a 2-*d* example is depicted. Then, the algorithm defines the next execution time for the *L-MM* module, namely the time an object will be deleted from the system.

2.2   The Eager Algorithm

The *Lazy* algorithm has some disadvantages: it stores obsolete data, i.e. tuples that will never be used in the skyline. This motivated its authors to consider the *Eager* algorithm [12], which aims to: (i) lower the memory consumption by keeping only the tuples that are or will be part of the skyline, and (ii) lower the cost of the maintenance module, in this case the *E-MM*. *Eager* achieves these two goals by doing more in the pre-processing module *E-PM*, where the *influence time* is computed to predict at arrival time, if a tuple will be included in a future skyline. If there is no such time, the object can safely be discarded. *Eager* uses an *Event List*, in which the events are sorted ascending based on the time of their respective events. Such events are the expiration of an object, or the transfer from the database to the skyline. Each tuple that is not part of the skyline but will be in the future, is *marked* and transferred to it at the proper time. Specifically in the *E-PM* module, for each incoming tuple, a query finds the tuples that are dominated by the incoming one. These tuples are then removed from the system. The new r tuple is inserted in the database and the *influence time* is computed by finding all the skyline objects that are in the $r.ADR$ region and then keeping the greatest expiry time. At that time point, tuple r will be transferred from the database to the skyline. If the computed influence time equals the arrival time, then the tuple is directly inserted in the skyline, whereas in the event list is marked with an *EX* value. Otherwise, it is stored in the database with an *EL* value.

When the time for an event comes, the *E-MM* method is executed. This method is less complicated than its respective in *Lazy* algorithm, because more processing is being done in the *E-PM* module. Thus, if the next event in the list is marked as *EX*, then the tuple is simply removed. Otherwise, the tuple is included in the skyline and a new event is stored in the event list to indicate the tuple expiry time.

3   The LookOut Algorithm

The *LookOut* algorithm connects each object with a time interval for which it is valid [8]. This time interval consists of the arrival time and the expiry time. The skyline can change in two occasions: (i) some skyline data are about to expire, and (ii) new data are inserted in the database.

The *LookOut* algorithm takes advantage of two important observations in hierarchical spatial indexes, e.g. R-trees [3] and quadtrees [10]: (i) if point p dominates all corners of a node n, then p dominates all the objects of the node and its children, and (ii) if all corners of a node n dominate a point p, then all objects and its children dominate that point. With these observations *pruning* of nodes is possible and rejection of new objects is faster. Each new object is inserted in the database and then the expiry time is stored in a min heap. The object is checked if it belongs to the skyline by an *isSkyline* algorithm. If an object must be removed, then all candidates that may replace it in the skyline are computed by a *MINI* algorithm. Final insertion is done only if *isSkyline* returns true.

isSkyline uses best-first search, i.e. nodes with the lowest distance are inserted first in the heap. When expanding a node, if the lower left corner does not dominate the arriving tuple, it is rejected. If the upper right corner of a child dominates the new tuple, then the algorithm terminates with negative output and the tuple is not inserted in the skyline. If the node is a leaf, the tuple is compared for domination against all tuples of the leaf. If there is such a leaf, the incoming object is not inserted in the skyline, otherwise it is.

The *MINI* algorithm uses also best-first search and a min heap for the distance from the origin to the point coordinates. An object that is about to be deleted is passed as an argument and returns the objects that are dominated by it. In addition, these objects are checked before insertion for domination by others that have already been inserted. According to the *isSkyline* algorithm, if the upper right corner is dominated by the object that is about to be deleted, the node is rejected, otherwise if it is an internal dominated node, it is inserted in the heap. If the node currently checked is a leaf, then the local skyline is computed and stored.

4   Experimentation and Evaluation

4.1   Methodology

For a thorough experimentation we rely on well selected datasets. For this reason three data types have been created, by using the generator of [1], and tested: (i) correlated data, (ii) anti-correlated data, and (iii) independent data. As spatial index we use R-trees [3, 7], which store objects in nodes dynamically generated at insertion time. Every node represents a *Minimum Bounding Rectangle* (MBR) and is created by using the coordinates of the lower-left and upper-right MBR corners. This way, during traversal it is possible to prune insignificant nodes.

The *Branch-and-Bound Skyline* (*BBS*) algorithm was used for all three algorithms [9]. The *BBS* algorithm traverses the tree and expands each node, storing in ascending order the distances from the axes origin. In each iteration, the node with the lowest distance is expanded or discarded. If the node is dominated by the existing skyline, it is rejected, otherwise kept. When the algorithm finds a leaf, it inserts the data in the skyline, because they already have been checked.

4.2   Improvement of Lazy Algorithm

According to the *Lazy* algorithm, the *EDR* must be computed for *L-MM* algorithm to work. This is easily achieved in 2-*d* datasets. An ascending ordered array is needed for each skyline dimension. Then finding the next value, after the point that is about to be deleted, creates a tuple for the upper-right corner of the *EDR*. The *EDR* region is computed by using the coordinates of the point to be deleted with the coordinates of the upper-right corner. This is depicted in Fig. 3.

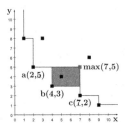

Fig. 3. Computing a 2-*d* EDR

However, in more than 2 dimensions the shape of the *EDR* becomes complicated and its computation hard or even impossible [14]. The authors of [12] do not clarify how the *EDRs* were computed and if the datasets allowed the creation of *EDRs* that could be computed easily as in 2-*d* datasets.

To correct these problems we made an improvement in the Lazy algorithm by replacing the *EDR* computation of a point with its dominance region (*DR*) for dimensions higher than 2. With this technique Lazy can return correct results for all dimensions and remain efficient.

4.3   Time Performance

Several tests were conducted by varying the dimensionality and the *SW* size on an Intel Core 2 Duo P8600, with 3GB RAM, 5400 RPM HDD and 64-bit Windows OS. In all tests the *Eager* algorithm prevails, as tuples are checked only once at arrival time if they belong in the skyline. In addition, the *Eager* algorithm has a linear scaling in all dimensions and *SW* sizes. The *Lazy* algorithm has similar performance for 2-*d* datasets; however, its performance is heavily compromised in higher dimensions. This is due to the dominance region in more than 2 dimensions. In this case, the search region is far greater than in the *EDR* region and, thus, the number of tuples to be checked is much greater as well.

The *LookOut* algorithm is worse in all cases. For small *SW* sizes the difference is comparable, but for sizes greater than some hundreds, the execution time increases dramatically. One reason is the *MINI* algorithm. For *mini-skyline* to be computed, all tuples that have not been pruned in the expansion phase, are potential insertions in the skyline and have to be checked. When the *SW* size gets larger, more tuples are possible members of the skyline and must be checked with each other. Another issue of the *LookOut* algorithm is after the execution of the *MINI*, when the *isSkyline* has to be executed, so that the potential members are sorted and accordingly rejected or inserted in the skyline.

In addition, all three algorithms seem to perform better in correlated data. This is probably due to better MBR creation and more effective pruning, which results in faster tree traversals. Tables 1, 2 and 3 contain the experimental results.

Table 1. Execution time (in seconds), for anti-correlated data

Dimensions	2-d			4-d			6-d		
SW size	Lazy	Eager	LookOut	Lazy	Eager	LookOut	Lazy	Eager	LookOut
100	16.05	6.17	24.32	14.19	6.49	102.87	29.91	7.09	184.71
1K	7.26	11.01	63.11	122.89	13.29	411.80	139.78	12.49	1264.03
10K	13.09	16.41	189.27	1249.90	30.09	1819.39	11415.70	27.59	9969.60

Table 2. Execution time (in seconds), for independent data

Dimensions	2-d			4-d			6-d		
SW size	Lazy	Eager	LookOut	Lazy	Eager	LookOut	Lazy	Eager	LookOut
100	5.06	5.44	21.56	14.18	7.56	113.59	25.24	6.72	162.12
1K	6.52	10.89	52.75	123.09	12.93	428.73	137.27	11.79	1231.35
10K	13.99	16.18	184.63	1776.30	31.15	1773.00	12522.34	47.05	12920.20

Table 3. Execution time (in seconds), for correlated data

Dimensions	2-d			4-d			6-d		
SW size	Lazy	Eager	LookOut	Lazy	Eager	LookOut	Lazy	Eager	LookOut
100	5.28	4.74	18.45	12.20	5.76	86.32	17.96	6.00	151.75
1K	5.48	9.03	45.29	103.70	11.62	323.01	117.59	10.93	1083.14
10K	7.12	14.85	166.96	1118.71	28.10	1624.72	11814.40	38.53	12492.90

4.4 Memory Consumption

Authors of [12] state that *Eager* algorithm was developed to consume less memory than *Lazy*. This is verified by the experiments, because even in the 6-d datasets and the largest *SW*, the algorithm consumes less than 10Mb of memory as shown in Table 4.

On the other hand, the *Lazy* and the *LookOut* algorithms have higher memory consumption, since they exceed in some cases dozens of Mb even in 2-d datasets. The *Lazy* algorithm displays fluctuations in the memory allocation at small *SW* sizes, but its memory consumption is linear in greater sizes. The *LookOut* algorithm has linear consumption, but when the *SW* size is 1M the memory consumption reaches and surpasses 50Mb (see Table 5).

Table 4. Memory consumption (in Mb) of *Eager* algorithms for 2-, 4-, 6-d data

SW size	2-d	4-d	6-d
100	0.5	0.5	0.5
1K	0.5	0.6	0.6
10K	0.5	0.8	1.2
100K	0.6	1.2	2.9
1M	0.6	1.2	6.9

Table 5. Memory consumption (in Mb) of *LookOut* and *Lazy* for 2-d data

SW size	Lazy	LookOut
100	10	0.5
1K	2.9	0.6
10K	2.3	1.4
100K	7.5	8.5
1M	67.5	58.0

5   Conclusions

This paper examines three skyline algorithms and compares their performance. Experiments established the fact that the dimensionality and the *SW* size are the main factors that affect the performance and the effectiveness of an algorithm, which is not clearly visible in small datasets. Also, the dominance region was used in the *Lazy* algorithm for the computation of the skyline, as the *Exclusive Dominance Region* is sometimes impossible to be computed in higher dimensions.

References

1. Börzsönyi, S., Kossmann, D., Stocker, K.: The skyline operator. In: Proceedings of the ICDE, pp. 421–430 (2001)
2. Chomicki, J., Godfrey, P., Gryz, J., Liang, D.: Skyline with presorting. In: Proceedings of the ICDE, pp. 717–719 (2003)
3. Guttman, A.: R-trees: A dynamic index structure for spatial searching. In: Proceedings of the SIGMOD, pp. 47–57 (1984)
4. Kossmann, D., Ramsak, F., Rost, S.: Shooting stars in the sky: an online algorithm for skyline queries. In: Proceedings of the VLDB, pp. 275–286 (2002)
5. Li, X., Wang, Y., Li, X., Wang, Y.: Parallel skyline queries over uncertain data streams in cloud computing environments. Int. J. Web Grid Serv. **10**(1), 24–53 (2014)
6. Lu, H., Zhou, Y., Haustad, J.: Efficient and scalable continuous skyline monitoring in two-tier streaming settings. Inf. Syst. **38**(1), 68–81 (2013)
7. Manolopoulos, Y., Nanopoulos, A., Papadopoulos, A., Theodoridis, Y.: R-Trees: Theory and Applications. Springer, Heidelberg (2005)

8. Morse, M., Patel, J., Grosky, W.: Efficient continuous skyline computation. Inf. Sci. **177**(17), 3411–3437 (2007)
9. Papadias, D., Tao, Y., Fu, G., Seeger, B.: An optimal and progressive algorithm for skyline queries. In: Proceedings of the SIGMOD, pp. 467–478 (2003)
10. Samet, H.: The quadtree and related hierarchical data structures. ACM Comput. Surv. **16**(2), 187–260 (1984)
11. Tan, K.L., Eng, P.K., Ooi, B.: Efficient progressive skyline computation. In: Proceedings of the VLDB, pp. 301–310 (2001)
12. Tao, Y., Papadias, D.: Maintaining sliding window skylines on data streams. TKDE **18**(3), 377–391 (2006)
13. Tzanakas, A., Tiakas, E., Manolopoulos, Y.: Revisited skyline query algorithms on streams of multidimensional data. Technical report (2016). http://delab.csd.auth.gr
14. Wu, P., Agrawal, D., Egecioglu, O., El Abbadi, A.: DeltaSky: optimal maintenance of skyline deletions without exclusive dominance region generation. In: Proceedings of the ICDE, pp. 486–495 (2007)
15. Xin, J., Wang, G., Chen, L., Zhang, X., Wang, Z.: Continuously maintaining sliding window skylines in a sensor network. In: Kotagiri, R., Radha Krishna, P., Mohania, M., Nantajeewarawat, E. (eds.) DASFAA 2007. LNCS, vol. 4443, pp. 509–521. Springer, Heidelberg (2007)

Canonical Data Model for Data Warehouse

Manuk Manukyan[1(✉)] and Grigor Gevorgyan[2]

[1] Yerevan State University, 0025 Yerevan, Armenia
mgm@ysu.am
[2] Russian-Armenian (Slavonic) University, 0051 Yerevan, Armenia
grigor.gevorgyan@gmail.com

Abstract. In the frame of our approach to data integration an extensible canonical model is developed. In the present paper we consider an extension of the canonical model to support materialized integration of data during data warehouse creation. The considered data warehouse is based on the concept of grid file. To extract data from several sources, to create the materialized view, and to effectively organize queries on the multidimensional data a computationally complete language is used.

Keywords: Data integration · Data warehouse · Mediator · Extensible canonical model · Grid file

1 Introduction

We have published a number of papers that are devoted to investigation of data integration problems (for instance see [17,18]). An extensible canonical model appeared as an outcome of this research. In the frame of this research an approach to virtual integration of data has been developed. In order to create the justifiable data model mapping for heterogeneous databases integration, the concept of data model has been formalized by means of the AMN (Abstract Machine Notation) [6] formalism. For each source model we create a reversible mapping into an extension of the canonical model. B-technology [6] is used to prove that the AMN semantics of the source model represents a refinement of the AMN semantics of the extended canonical model. Hereby the correctness of mapping and the ability to use extended canonical model for representation of the source model schemas are proved. AMN-machines for the canonical and relational data models and relational refinement of the canonical data model have been created. The theoretical basis of our approach to data integration are the works of the SYNTHESIS group (IPI RAS) [12–14], who are pioneers in the area of justifiable data models mapping for heterogeneous databases integration. According to this approach, each data model is defined by syntax and semantics of two languages, data definition language (DDL) and data manipulation language (DML). The main principle of mapping of an arbitrary resource data model into the target

This work was supported by the RA MES State Committee of Science, in the frames of the research project N 15T-18350.

M. Ivanović et al. (Eds.): ADBIS 2016, CCIS 637, pp. 72–79, 2016.
DOI: 10.1007/978-3-319-44066-8_8

one (the canonical model) could be reached under the condition, that the diagram of DDL (schemas) mapping and the diagram of DML (operators) mapping are commutative.

In the present paper an extension of the canonical model is considered to support materialized integration of data during creation of a data warehouse. The canonical model kernel is a result of minor extension of the OPENMath (a standard for the representation and communication of mathematical objects) [8] for supporting the concept of databases. Choice of OpenMath as the basic formalism for canonical model allows to use a rich mathematical apparatus during data warehouse creation and to apply very complex OLAP-queries on it.

The paper is organized as follows: the principles of canonical data model construction are considered briefly in Sect. 2. Formalizations of the grid file and data integration concepts by means of XML are proposed in Sects. 3 and 4 correspondingly. Related work is presented in Sect. 5. The conclusion is provided in Sect. 6.

2   Extensible Canonical Model

The canonical model kernel is an ontological model which we use to formalize concepts of data models. Within this model we distinguish basic and compound objects. A data model concept formalized on the kernel level is referred to as a *kernel object*. Formally, a kernel object is a labeled tree whose leaves are basic kernel objects. Examples of basic kernel objects are constants, variables, and symbols (for instance, reserved words). The compound objects are defined in terms of *binding* and *application* of λ-calculus [11]. The type system is built on the basis of types that are defined by themselves and certain recursive rules, whereby the compound types are built from simpler types. The basis consists of the conventional atomic types (for example, *integer*, *string*, *boolean*, etc.). We use the following type constructors to build compound types:

- *Attribution.* If v is a basic object variable and t is a typed object, then **attribution**(v, *type t*) is typed object. It denotes a variable with type t.
- *Abstraction.* If v is a basic object variable and t, A are typed objects, then **binding**(*lambda*, **attribution**(v, *type t*), A) is typed object.
- *Application.* If F and A are typed objects, then **application**(F, A) is typed object.

Addition of applicative syntax and semantics substantially increases the expressiveness of the XML data model. For this data model a declarative query language (element calculus)[1] is developed [16]. We can combine the possibility of the applicative programming with element calculus to give very complex *integrity constraints* on the global schema level. The necessity to give such integrity constraints arises since the data of the sources may not satisfy the global schema constraints.

---

[1] The detailed discussion of the canonical model query language is beyond the topic of this paper.

2.1 Semantic Level

The canonical model kernel is implemented as an XML application. Its syntax is defined by syntactical rules of XML, its grammar is partially defined by its own DTD. Only syntactical validity of the kernel objects representation can be provided on the DTD level. To check semantics, in addition to general rules inherited by XML applications, the considered application defines new syntactical rules. This is achieved by means of introduction of *content dictionaries*. Content dictionaries are used to assign formal and informal semantics to all symbols (for example, *type*, *sequence*, *OneOrMore*, *string*, etc.) used in the kernel objects. A content dictionary is a collection of related symbols encoded in XML format. In other words, each content dictonary defines symbols representing a concept from the specific subject domain.

2.2 Kernel Extension Principles

The canonical model must be extensible [17]. The extension of the canonical model is formed during consideration of each new data model by adding new concept(s) to its DDL to define logical data dependencies of the source model in terms of the target model if necessary. Thus, the kernel extension assumes defining new symbols. The extension result must be equivalent to the source data model. For applying a *symbol* on the canonical model level the following rule is proposed:

Concept ← *symbol* ContextDefinition

For example, to support the concepts of *key* of relational data model, we have expanded the kernel with the symbol *key*. Let us consider a relational schema example: S = {Snumber, Sname, Status, City}. The equivalent definition of this schema by means of extended kernel is considered below:

S ← *attribution*(S, *type* TypeContext, *constraint* ConstraintContext)

TypeContext ← *application*(*sequence*, ApplicationContext)

ApplicationContext ← *attribution*(Snumber, *type int*),

                 attribution(Sname, *type string*), *attribution*(Status,

                 type int), *attribution*(City, *type string*)

ConstraintContext ← *attribution*(ConstraintName, *key* Snumber)

It is essential that we use a computationally complete language to define the context [16]. As a result of such approach usage of new symbols in the DDL does not lead to any changes in DDL parser. According to this approach, the canonical model is synthesized as a union of extensions. A schema of the integrated databases is an instance of the canonical model.

3 Formalization of the Grid File Concept

The emergence of a new paradigm in science and various applications of information technology (IT) is related to issues of big data handling [22]. This concept involves the growing role of data in all areas of human activity beginning with research and ending with innovative developments in business. Such data is difficult to process and analyze using conventional database technologies. In this connection, the creation of new IT is expected in which data becomes dominant for new approaches to conceptualization, organization, and implementation of systems to solve problems that were previously considered extremely hard or, in some cases, impossible to solve. Unprecedented scale of development in the big data area and the U.S. and European programs related to big data underscore the importance of this trend in IT. The concept of grid files [19, 20] is one of the adequate formalisms for effective big data management. In fact, the concept of grid files allows to effectively organize queries on multidimensional data [9] and can be used for efficient data cubes storage in data warehouses [15, 20].

We developed a new dynamic index structure for multidimensional data [10]. The considered index structure is based on the grid file concept. This concept has been extended and formalized by means of XML. Efficient algorithms for storage and access of the index directory are developed, in order to minimize memory usage and lookup operations complexities. Estimations of complexities for these algorithms are obtained. The grid file can be represented as if the space of points is partitioned into an imaginary grid. The *grid lines* parallel to axis of each dimension divide the space into *stripes*. The number of grid lines in different dimensions may vary, and there may be different spacings between adjacent grid lines, even between lines in the same dimension.

One of the problems intrinsic to grid files is the problem of non-efficient memory usage by groups of cells referring to the same data buckets. We introduce an effective dynamic *chunking* scheme – a partitioning technique for directory in order to create united address for some information in the cube, i.e., the so-called *chunks*. Firstly, we refuse from storing the grid file as a multidimensional array. The reason for that lies within the necessity of creation of numerous new cells each time a data bucket is split, while many of those cells contain duplicate pointers to the same data buckets. Instead, all cells which contain pointers to the same data buckets are grouped into chunks, represented by single memory cells with one pointer to the considered data bucket. Chunks are the main units for data I/O, as well as are used for data clusterisation. Chunks are also used as a mechanism of struggle with empty cells problem in grid file. For each dimension the information about its division is stored in a linear scale, each element of which corresponds to a stripe of the grid file and is represented as an array of pointers to the chunks, crossed by that stripe.

Secondly, we consider each stripe as a linear hash table, allowing usage of overflow blocks to reduce the number of chunk partitions. The number of overflow blocks may be different for different chunks, however we ensure that for any stripe the average number of overflow blocks for the chunks crossed by that stripe is not greater than one. This allows us to significantly reduce the total amount of chunks, while guaranteeing not more than two disk operations for data access in average.

The aim of XML formalization of the grid file concept is to create a directory definition language which will be independent of data management paradigms. Formalization of the grid file concept by means of XML assumes an XML application development. In this application we define grid file concepts of dimension, stripe and chunk as XML elements.

Element *stripe* is an empty element and is described by means of five attributes: *ref_to_chunk*, *min_val*, *max_val*, *rec_cnt* and *chunk_cnt*. The values of attributes *ref_to_chunk* are pointers to chunks crossed by each stripe. By means of *min_val* (lower boundary) and *max_val* (upper boundary) attributes we define "widths" of the stripes. The values of attributes *rec_cnt* and *chunk_cnt* are the total number of records in a stripe and the number of chunks crossed by it correspondingly.

The content of element *chunk* is based on the attributes *avg*, *id* of type *ID*, *ref_to_db* and *ref_to_chunk*. Values of attributes *ref_to_db* and *ref_to_chunk* are pointers to data blocks and other chunks correspondingly. Values of *avg* attributes are used during reorganization of the grid file and contain the average coordinates of points, corresponding to records of the considered chunk, for each dimension.

The content of element *dim* is based on its inner elements *stripe* and has a single attribute *name* - the dimension name.

4   Formalization of the Data Integration Concept

In this section we propose an approach to data integration which is based on the ontology. In the frame of this approach we consider virtual as well as materialized data integration issues whitin a canonical model. Therefore, we should formalize the concepts of this subject area such as mediator, data warehouse and database schema. In our case the data integration ontology is based on the XML representation of these concepts. We are modelling these concepts by means of XML elements *dbsch*, *med* and *whse* correspondingly.

The content of element *dbsch* is based on the kernel element *omattr*. By means of *omattr* element we can model schemas of databases (more details see Sect. 2).

The content of element *med* is based on the elements *msch*, *wrapper*, *constraint* and has an attribute *name*. The value of this attribute is the mediator's name. The element *msch* is interpreted analogously to element *dbsch*. Only note that this element is used during modelling schemas of mediator. The content of elements *wrapper* and *constraint* is based on the kernel element *omobj* (mathematical object, more details see [8]). By means of *wrapper* element mappings from source models into canonical model are defined. The integrity constraints on the level of mediator are the values of the *constraints* elements.

Our approach to support data warehousing is based on the grid file concept and is interpreted by means of element *whse* which contains elements *wsch*, *extractor*, *grid* and has an attribute *name*. The value of this attribute is the name of data warehouse. The elements *wsch* and *extractor* are interpreted in the same way as the elements *msch* and *wrapper* for the mediator. The element *grid* contains elements *dim* and *chunk* by which the grid file concept is modelled (for more details see previous section).

5   Related Work

There is a variety of approaches to big data management today (for example, [1–5,7,21,23,24]). Detailed analysis of big data systems can be found in [22]. In this section we shall compare one of them (namely MongoDB [3,4]) with our approach in the context to support big data. MongoDB is an open-source document-oriented No-SQL database, providing high performance, availability and easy scalability. MongoDB supports different types of indices, namely: single field, compound, multikey, geospatial, text and hashed indices. MongoDB indices are implemented as B-Trees. We have implemented a data warehouse prototype based on the proposed dynamic indexation scheme and compared its performance with MongoDB. The latter was choosen for comparison for pragmatical reasons since it is currently one of the most demanded NoSQL databases.

Testing was conducted using four main query categories [9] - given point lookup, lookup by individual coordinates, range lookup and closest object lookup. Comparison of directory sizes depending on the number of database records was also performed. Several results of tests are presented below. We used points in 3-dimensional Euclidean space for these tests, where points coordinates were represented with 32-bit unsigned integers.

Figure 1(a) presents charts of index directory size growth when performing 2 mln. insert operations. Figure 1(b) presents charts, displaying the time required to process 10^5 given point lookup operations depending on the number of records in database. Here we can see that our data warehouse prototype index structure requires much less memory than B-trees used by MongoDB and processes point lookup queries faster. A query for data range lookup defineds a rectangular area of the grid, and answer to it is composed by set of points belonging to the chunks which are crossed by that area.

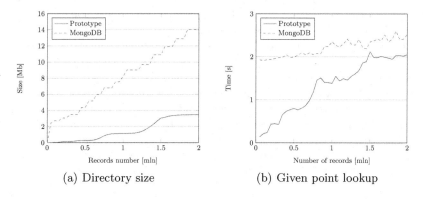

(a) Directory size   (b) Given point lookup

Fig. 1. Directory size and given point lookup

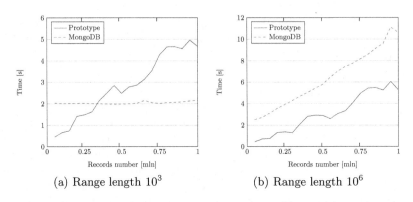

(a) Range length 10^3 (b) Range length 10^6

Fig. 2. Range lookup

Figure 2 contains comparison of range queries performed in 3-dimensional space with values uniformly distributed in range $[0..2^{32} - 1]$ and average range length of 10^3 and 10^6 correspondingly. It should be noted that with the increase of query range length our prototype performs better in comparison with MongoDB. In general, the conducted experiments show that our data warehouse prototype performs faster during given point lookup, lookup in wide data ranges, closest point lookup, and also provides more effective memory usage for index directory.

6   Conclusions

In the present paper we are considering issues of materialized integration of data. In the capacity of integrating model we use a canonical data model which is a result of our research within an approach to data integration. The offered data warehouse is based on the a new dynamic indexing structure for multi-dimensional data. The latter is an effective indexing structure and has been developed by us to support our approach to data integration. The considered ontology of data integration is based on the XML representation of basic concepts of mediator and warehouse data integration technologies. It is essential that we use a computational complete language for integrating data. Namely, to create schemas of mediator and warehouse as well to give queries to mediator and extract data for warehouse from data sources. In order to use our approach to data integration we have implemented a data warehouse prototype based on the proposed dynamic indexation scheme and compared its performance with MongoDB. The considered warehousing technology can also be used in the context of big data. Finally, our approach to materialized integration of data allows to combine the applicative programming possibilities with element calculus and to give very complex OLAP-queries to data warehouse.

References

1. http://cassandra.apache.com
2. http://neo4j.com
3. http://paradigm4.com
4. https://www.mongodb.org
5. http://www.couchbase.com
6. Abrial, J.-R.: The B-Book-Assigning Programs to Meaning. Cambridge University Press, Great Britain (1996)
7. Chang, F., Dean, J., Ghemawat, S., Hsieh, W., Wallach, D.A., Burrows, M., Chandra, T., Fikes, A., Gruber, R.E.: Bigtable: a distributed storage system for structured data. In: OSDI 2006, November 2006
8. Drawar, M.: Openmath: an overview. ACM SIGSAM Bull. **34**(2), 2–5 (2000)
9. Garcia-Molina, H., Ullman, J., Widom, J.: Database Systems: The Complete Book. Prentice Hall, USA (2009)
10. Gevorgyan, G.R., Manukyan, M.G.: Effective algorithms to support grid files. RAU Bull. **2**, 22–38 (2015)
11. Hindley, J.R., Seldin, J.P.: Introduction to Combinators and λ-Calculus. Cambridge University Press, Great Britain (1986)
12. Kalinichenko, L.A.: Data model transformation method based on axiomatic data model extension. In: 4th International Conference on VLDB, pp. 549–555. Springer, Germany, September 1978
13. Kalinichenko, L.A.: Methods and Tools for Integration of Heterogeneous Databases (in Russian). Science, USSR (1983)
14. Kalinichenko, L.A.: Methods and tools for equivalent data model mapping construction. In: Bancilhon, F., Zhang, J., Thanos, C. (eds.) EDBT 1990. LNCS, vol. 416, pp. 92–119. Springer, Heidelberg (1990)
15. Luo, C., Hou, W.C., Wang, C.F., Want, H., Yu, X.: Grid file for efficient data cube storage. Computers and their Applications, pp. 424–429 (2006)
16. Manukyan, M.G.: Extensible data model. In: Advances in Databases and Information Systems, pp. 42–57. Finland (2008)
17. Manukyan, M.G.: Canonical model: construction principles. In: iiWAS2014, pp. 320–329. ACM, Vietnam, December 2014
18. Manukyan, M.G., Gevorgyan, G.R.: An approach to information integration based on the amn formalism. In: First Workshop on Programming the Semantic Web, USA, pp. 1–13, November 2012. https://web.archive.org/web/20121226215425/, http://www.inf.puc-rio.br/~psw12/program.html
19. Nievergelt, J., Hinterberger, H.: The grid file: an adaptable, symmetric, multikey file structure. ACM Trans. Database Syst. **9**(1), 38–71 (1984)
20. Papadopoulos, A.N., Manolopoulos, Y., Theodoridis, Y., Tsoras, V.: Grid file (and family). In: Encyclopedia of Database Systems, pp. 1279–1282 (2009)
21. Robinson, I., Webber, J., Eifrem, E.: Graph Databases. O'Reilly, USA (2015)
22. Sharma, S., Tim, U.S., Wong, J., Gadia, S., Sharma, S.: A brief review on leading big data models. Data Sci. J. **13**, 138–157 (2014)
23. Stonebraker, M., Brown, P., Poliakov, A., Raman, S.: The architecture of SciDB. In: Bayard Cushing, J., French, J., Bowers, S. (eds.) SSDBM 2011. LNCS, vol. 6809, pp. 1–16. Springer, Heidelberg (2011)
24. The SciDB Development Team http://www.scidb.org. Overview of scidb. In: SIGMOD 2010, June 2010

Rhone: A Quality-Based Query Rewriting Algorithm for Data Integration

Daniel A.S. Carvalho[1], Plácido A. Souza Neto[2(✉)], Chirine Ghedira-Guegan[3], Nadia Bennani[4], and Genoveva Vargas-Solar[5]

[1] Université Jean Moulin Lyon 3, Centre de Recherche Magellan, IAE, Lyon, France
`daniel.carvalho@univ-lyon3.fr`
[2] Instituto Federal do Rio Grande do Norte - IFRN, Natal, Brazil
`placido.neto@ifrn.edu.br`
[3] Université de Lyon, CNRS, IAE - Université Lyon 3,
LIRIS, UMR5205, Lyon, France
`chirine.ghedira-guegan@univ-lyon3.fr`
[4] LIRIS, INSA-Lyon, Villeurbanne, France
`nadia.bennani@insa-lyon.fr`
[5] CNRS-LIG, Grenoble, France
`genoveva.vargas@imag.fr`

Abstract. Nowadays, data provision is mostly done by data services. Data integration can be seen as composition of data services and data processing services that can deal with to integrate data collections. With the advent of cloud, producing service compositions is computationally costly. Furthermore, executing them can require a considerable amount of memory, storage and computing resources that can be provided by clouds. Our research focuses on how to enhance the results on the increase of cost on data integration in the new context of cloud. To do so, we present in this paper our original data integration approach which takes into account user's integration requirements while producing and delivering the results. The service selection and service composition are guided by the service level agreement - SLA exported by different services (from one or more clouds) and used by our matching algorithm (called *Rhone*) that addresses the query rewriting for data integration presented here as proof of concept.

Keywords: Data integration · Query rewriting · Query rewriting algorithm · Cloud computing · SLA

1 Introduction

Current data integration implies consuming data from different data services and integrating the results according to different quality requirements related to data cost, provenance, privacy, reliability, availability, among others. Data services and data processing services can take advantage from the on-demand and *pay-as-you-go* model offered by the cloud architecture. The quality conditions and

© Springer International Publishing Switzerland 2016
M. Ivanović et al. (Eds.): ADBIS 2016, CCIS 637, pp. 80–87, 2016.
DOI: 10.1007/978-3-319-44066-8_9

penalties under which these services are delivered can be defined in contracts using Service Level Agreements (SLA).

Cloud services (data services, data processing services, for instance) and the cloud provider export their SLA specifying the level of services the user can expect from them. A user willing to integrate data establishes a contract with the cloud provider guided by an economic model that defines the services he/she can access, the conditions in which they can be accessed (duplication, geographical location) and their associated cost. Thus, for a given requirement, cloud services (from one or several cloud providers) are chosen to retrieve, process and integrate data according to the type of contracts he/she established with them.

In this context, data integration deals with a matching problem of the user's integration preferences which includes quality constraints and data requirements, and his/her specific cloud subscription with the SLA's provided by cloud services. Matching SLAs can imply dealing with heterogeneous SLA specifications and SLA-preferences incompatibilities. Moreover, even with the possibility of having an unlimited access to cloud resources, the user is limited to the resources and to the budget agreed by his/her cloud subscription. Inspired by these problems and carrying on the ideas presented in our previous work [3], the aim of this paper is to introduce our service-based query rewriting algorithm guided by user preferences and SLAs which enhances the quality on the results integration in a multi-cloud context.

This paper is organized as follows. Section 2 discusses related works. Section 3 describes the *Rhone* algorithm and its formalization. Experiments and results are described in the Sect. 4. Finally, Sect. 5 concludes the paper and discusses future works.

2   Related Works

In recent years, the cloud have been the most popular deployment environment for data integration [5]. Researches have proposed their works addressing this issue [6,8]. Moreover, once query rewriting is strictly related to data integration, rewriting algorithms have been presented [1,2,4].

In [6], the authors introduced a system (called SODIM) which combines data integration, service-oriented architecture and distributed processing. The novelty of their approach is that they perform data integration in service oriented contexts, particularly considering data services. A major concern when integrating data from different sources (services) is privacy that can be associated to the conditions in which integrated data collections are built and shared. [8] proposed an inter-cloud data integration system that considers a trade-off between users' privacy requirements and the cost for protecting and processing data. According to the users' privacy requirements, the query plan in the cloud repository creates the users' query. Thus, the query plan executor decides the best location to execute the sub-query to meet privacy and cost constraints. This work is mostly interested in privacy and performance issues forgetting other users' integration requirements.

The main aspect in a data integration solution is the query rewriting. In the database domain, the query rewriting problem using views have been widely discussed [7]. Similarly, data integration can be seen in the service-oriented domain as a service composition problem in which given a query the objective is to lookup and compose data services that can contribute to produce a result. Generally, data integration solutions on the service-oriented domain deal with query rewriting problems. [2] proposed a query rewriting approach which processes queries on data provider services. [4] introduced a service composition framework to answer preference queries. Two algorithms are presented to rank the best rewritings based on previously computed scores. [1] presented an refinement algorithm based on *MiniCon* that produces and order rewritings according to user preferences and scores used to rank services that should be previously define by the user. Furthermore, they do not take into consideration user's integration requirements which can lead to produce rewritings that are not satisfactory in terms of quality requirements and constraints imposed by the user and the cloud environment. We assume that these requirements and constraints be expressed on SLAs. In the next section, we introduce our query rewriting algorithm that deals with SLAs while selecting, filtering and producing results.

3   Rhone Service-Based Query Rewriting Algorithm

This section describes our quality-based query rewriting algorithm called *Rhone*. It is guided by user requirements and constraints, and services' quality features extracted after structuring service level agreements (SLA). Our algorithm has two original aspects: *first*, the user can express his/her quality requirements and constraints, and associate them to his/her queries; and, *second*, service's quality features defined on Service Level Agreements (SLA) guide service selection and the whole rewriting process.

3.1   Preliminaries

The idea behind our algorithm consists in deriving a set of service compositions that fulfill the users' integration preferences and requirements concerning the context of data service deployment given a set of *abstract services*, a set of *concrete services*, a user' *query* defined hereafter and a set of user' *integration preferences and requirements*.

Definition 1 (*Abstract service*). An *abstract service* describes the small piece of function that can be performed by a cloud service. For instance, retrieve patients, DNA information and personal information. The *abstract service* is defined as $A\,(\overline{I};\,\overline{O})$ where: A is a name which identifies the *abstract service*; and $\overline{I}$ and $\overline{O}$ are a set of comma-separated input and output parameters, respectively. The Table 1 exemplifies *abstract services* in a medical scenario. The decorations ? and ! differentiate input and output parameters, respectively.

Table 1. List of *abstract services.*

Abstract service name	Description
A1 (*d_name*?; *p_id*!)	Given a disease name *d_name*, A1 returns a list of infected patients' id *p_id*
A2 (*p_id*?; *p_dna*!)	Given a patient id *p_id*, A2 returns her DNA information *p_dna*
A3 (*p_id*?; *p_info*!)	Given a patient id *p_id*, A3 returns her personal information *p_info*
A4 (*d_name*?; *regions*!)	Given a disease name *d_name*, A4 returns the most affected region *regions*

Definition 2 (*Concrete service*). A *concrete service* is defined as a set of *abstract services,* and by its *quality features* extracted from its SLA according to the following grammar:

$$S(\overline{I}_h; \overline{O}_h) := A_1(\overline{I}_{1l}; \overline{O}_{1l}), A_2(\overline{I}_{2l}; \overline{O}_{2l}), .., A_f(\overline{I}_{fl}; \overline{O}_{fl})[M_1, M_2, .., M_g]$$

The left-hand of the definition is called the *head*; and the right-hand is the *body*. A *concrete service* S includes a set of input $\overline{I}$ and output $\overline{O}$ variables, respectively. Variables in the *head* are identified by $\overline{I}_h$ and $\overline{O}_h$, and called *head* variables. They appear in the *head* and in the *body* definition. Variables appearing only in the *body* are identified by $\overline{I}_l$ and $\overline{O}_l$, and are called *local* variables. *Head* variables can be accessed and shared among different services. On the other hand, *local* variables can be used only by the service which define them.

Concrete services are defined in terms of *abstract services* $(A_1, A_2, .., A_n)$, and they include a set of service's quality features $(M_1, M_2, .., M_g)$ that are extracted from the SLA exported by the service S. M_i is in the form $x \otimes c$, where x is a special class of identifiers associated to the services; c is a constant; and $\otimes \in \{\geq, \leq, =, \neq, <, >\}$.

Let us consider the following *concrete services* specified using the *abstract services* previously presented to be used as examples to the algorithm:

```
S1(a?;b!) := A1(a?;b!) [availability > 98%, price per call = 0.2$]
S2(a?;b!) := A1(a?;b!) [availability > 98%, price per call = 0.1$]
S3(a?;b!) := A2(a?;b!) [availability > 99%, price per call = 0.1$]
S4(a?;b!) := A1(a?;p!), A2(p?; b!)
                        [availability > 98%, price per call = 0.1$]
S5(a?;b!) := A3(a?;b!) [availability > 98%, price per call = 0.0$]
S6(a?;b!,c!) := A1(a?;p!), A2(p?;b!), A3(p?;c!)
                        [availability > 99%, price per call = 0.2$]
S7(a?;b!) := A4(a?;b!) [availability > 99%, price per call = 0.2$]
```

For instance, $S1$ is written using the *abstract service* $A1$. a and b are *head* variables. *Availability* and *price per call* are identifiers associated to $S1$'s quality features with an associated constant value extracted from $S1$'s SLA.

Definition 3 (*User query*). A user *query* Q is defined as a set of *abstract services*, a set of *constraints*, and a set of *user integration requirements* in accordance with the following grammar:

$$Q(\overline{I}_h; \overline{O}_h) := A_1(\overline{I}_{1l}; \overline{O}_{1l}), A_2(\overline{I}_{2l}; \overline{O}_{2l}), .., A_n(\overline{I}_{nl}; \overline{O}_{nl}), C_1, C_2, .., C_m[P_1, P_2, .., P_k]$$

The *query* definition is similar to a *concrete service* concerning the variables and *abstract services*. In addition, queries have constraints over the input or output variables ($C_1, C_2, .., C_m$). Constraints are used while querying the databases (*i.e* in the "*where*" clause). The user *integration requirements* over the services or over service compositions are specified in $P_1, P_2, .., P_k$. C_i and P_j are in the form $x \otimes c$, where x is an identifier; c is a constant; and $\otimes \in \{\geq, \leq, =, \neq, <, >\}$.

User requirements can be of two types, single and composed. Single are associated directly to each service involved in the composition. Composed are linked to the entire composition. They are defined in terms of single requirements. For instance, the total response time is a composed preference obtained by adding the response time of each service involved in the composition.

Let us suppose a query specification based on a medical scenario in which doctor *Marcel* wants to query the personal and DNA information from patients that were infected by *flu*, using services with availability higher than 98 %, price per call less than 0.2\$ and integration total cost less than 5\$. To achieve his needs, the *abstract services* A1, A2 and A3 should be composed as follows.

```
Q(dis?;dna!,info!) := A1(dis?;p!), A2(p?;dna!), A3(p?;info!), d ="flu",
                      [availability > 98%, price per call < 0.2$, total cost < 5$]
```

The *Marcel's query* plan begins by retrieving infected patients (*A1*). This operation returns patients' ids p. The *abstract services* A2 and A3 use patient ids to return their DNA and personal information (*dna* and *info*). The *query* contains a constraint *dis* (disease name) equal to *flu*, and three identifiers define the user' *integration preferences* with their associated constant value.

The input data for the *Rhone* is a query and a set of concrete services. The result is a set of rewriting of the query in terms of concrete services, fulfilling the user preferences. The main function of the algorithm is divided in four steps: selecting candidate concrete services, creating candidate service descriptions and combining and producing rewritings. In the next sections, each step of the algorithm will be described.

3.2   Selecting Candidate Concrete Services

While selecting services, the algorithm deals with three matching problems: (*i*) *quality features* matching, each *feature* in a query should be found in a concrete service. Moreover, the evaluation of a *feature* in a concrete service must satisfy the evaluation of a *feature* in the query; (*ii*) *abstract service* matching, abstract services can be matched if they have the same abstract function name and if the number and type of variable are equivalent; and (*iii*) *concrete service* matching, all abstract services in the concrete service must exist in the query, and all of them should satisfy the *feature* and *abstract service matching* problems. Compared to [1], our algorithm includes the *features* matching and extends the *concrete service* matching by not accepting *concrete services* that covers useless *abstract services* to the query rewriting.

3.3   Candidate Service Description Creation

After producing the set of candidate concrete services, the next step creates candidate service descriptions (CSDs). A CSD maps abstract services and variables of a concrete service into abstract services and variables of the query.

Definition 4 (candidate service description). A CSD is represented by an n-tuple:

$$\langle S, h, \varphi, G, P \rangle$$

where S is a *concrete service*. h are mappings between variables in the *head* of S to variables in the *body* of S. φ are mapping between variables in the *concrete service* to variables in the *query*. G is a set of *abstract services* covered by S. P is a set of *quality features* associated to the service S.

A CSD is created according to 4 rules: (i) for all head variables in a concrete service, the mapping h from the head to the body definition must exist; (ii) Head variables in concrete services can be mapped to head or local variables in the query; (iii) Local variables in concrete services can be mapped to head variables in the query; and (iv) Local variables in concrete services can be mapped to local variables in the query if and only if the concrete service covers all abstract services in the query that depend on this variable. The relation "depends" means that an output local variable is used as input in another abstract service.

Given the query Q and a list of candidate concrete services $\mathcal{L}_S$, a list of CSDs $\mathcal{L}_{CSD}$ is produced. A CSD is created only for candidate concrete services in which the mappings rules are being satisfied.

Given the candidate concrete services S2, S3, S4 and S5 selected in the previous step. The algorithm builds CSDs to S2, S3 and S5 once they satisfy all the mapping rules as follows. For instance, CSD_2 is produced to S2 as follows:
$\langle S2,\ h = \{a \to a,\ b \to b\},\ \varphi = \{a \to dis,\ b \to p\},\ G = \{A1\},\ P = \{availability > 98\%,\ price\ per\ call = 0.1\$\}\rangle$.

However, a CSD for S4 is not build because it violates the rule for local variables. It contains a local variable (p) mapped to a local variable in the query. Consequently, S4 must cover all abstract services in the query depending on this variable, but the abstract service $A3$ is not covered.

3.4   Combining and Producing Rewritings Step

Given the list of CSDs $\mathcal{L}_{CSD}$ produced, the *Rhone* produces all possible combinations of its elements. Building combinations deals with a NP hard complexity problem. The effort to process combinations increases while the number of CSDs and abstract services in the query increases.

The last step identifies rewritings matching with the query and fulfilling the user preferences. The *Rhone* algorithm verifies if a given CSD list p is a rewriting of the original query. The function return *true* if (i) the number of abstract services resulting from the union of all CSDs in p is equal to the number of abstract services in the query; and (ii) the intersection of all abstract services in each CSD on p is empty. It means that is forbidden to have abstract services replicated among the set p.

Let us consider CSD_2, CSD_3 and CSD_5 are CSDs that refer to the concrete services S2, S3 and S5, respectively. The *Rhone* produces combinations taking into account the part of the query covered by the service as follows:

$$p_1 = \{CSD_2\}$$
$$p_2 = \{CSD_2, CSD_3\}$$
$$p_3 = \{CSD_2, CSD_3, CSD_5\}$$

Given the combinations, the *Rhone* checks if each one of them is a valid rewriting of the original query.

- p_1 and p_2 are not valid rewritings; their number of abstract services do not match with the number of abstract services in the query.
- p_3 is a valid rewriting; the number of abstract services matches and there is no repeated abstract service.

4   Evaluation

Different experiments were produced to analyze the algorithm's behavior. The *Rhone*[1], so far, was evaluated in a local controlled environment simulating a mono-cloud including a service registry of 100 concrete services. Some experiments were produced to analyze the its behavior concerning performance, and quality and cost of the integration. The experiments include two different approaches: (i) the *traditional approach* in which user preferences and SLAs are not considered; and (ii) the *preference-guided approach* (P-GA) which considers the users' integration requirements and SLAs. Figures 1a and b summarize our first results.

The results P-GA are promisingly. Our approach increases performance reducing rewriting number which allows to go straightforward to the rewriting solutions that are satisfactory avoiding any further backtrack and thus reducing successful integration time (Fig. 1a). Moreover, using the P-GA to meet the user preferences, the quality of the rewritings produced has been enhanced and the integration economic cost has considerable reduced while delivering the expected results (Fig. 1b).

---

[1] The *Rhone* algorithm is implemented in Java and it includes 15 java classes in which 14 of them model the basic concepts (*query, abstract services, concrete services*, etc.), and 1 responsible to implement the core of the algorithm.

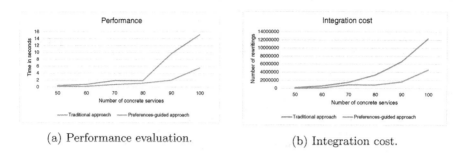

(a) Performance evaluation.   (b) Integration cost.

Fig. 1. *Rhone* execution evaluation.

5   Final Remarks and Future Works

Rhone still need to be tested in a large scale case and in a context of parallel multi-tenant to test efficacy. However, the results can show that the *Rhone* reduces the rewriting number and processing time while considering user preferences and services' quality aspects extracted from SLAs to guide the service selection and rewriting. We are currently performing a multi-cloud simulation in order to evaluate the performance of the *Rhone* in such context.

References

1. Ba, C., Costa, U., Halfeld-Ferrari, M., Ferre, R., Musicante, M.A., Peralta, V., Robert, S.: Preference-driven refinement of service compositions. In: International Conference on Cloud Computing and Services Science, Proceedings of CLOSER 2014 (2014)
2. Barhamgi, M., Benslimane, D., Medjahed, B.: A query rewriting approach for web service composition. IEEE Trans. Serv. Comput. **3**, 206–222 (2010)
3. Bennani, N., Ghedira-Guegan, C., Musicante, M., Vargas-Solar, G.: Sla-guided data integration on cloud environments. In: 2014 IEEE 7th International Conference on Cloud Computing (CLOUD), pp. 934–935, June 2014
4. Benouaret, K., Benslimane, D., Hadjali, A., Barhamgi, M.: FuDoCS: a web service composition system based on fuzzy dominance for preference query answering. In: 37th International Conference on Very Large Data Bases (VLDB 2011) (2011)
5. Carvalho, D.A.S., Neto, P.A.S., Vargas-Solar, G., Bennani, N., Ghedira, C.: Can data integration quality be enhanced on multi-cloud using SLA? In: Chen, Q., Hameurlain, A., Toumani, F., Wagner, R., Decker, H. (eds.) DEXA 2015. LNCS, vol. 9262, pp. 145–152. Springer, Heidelberg (2015)
6. ElSheikh, G., ElNainay, M.Y., ElShehaby, S., Abougabal, M.S.: SODIM: service oriented data integration based on MapReduce. Alexandria Eng. J. **52**, 313–318 (2013)
7. Halevy, A.Y.: Answering queries using views: a survey. VLDB J. **10**(4), 270–294 (2001)
8. Tian, Y., Song, B., Park, J., Huh, E.-N.: Inter-cloud data integration system considering privacy and cost. In: Pan, J.-S., Chen, S.-M., Nguyen, N.T. (eds.) ICCCI 2010, Part I. LNCS, vol. 6421, pp. 195–204. Springer, Heidelberg (2010)

Towards Spatial Crowdsourcing
in Vehicular Networks Using Mobile Agents

Oscar Urra$^{(\boxtimes)}$ and Sergio Ilarri

Department of Computer Science and Systems Engineering,
University of Zaragoza, I3A, Zaragoza, Spain
ourra@itainnova.es, silarri@unizar.es

Abstract. In the last years, the automotive industry has shown interest
in the addition of computing and communication devices to cars, thanks
to the technological advances in these fields but also to meet the increasing demand of "connected" applications and services. Although vehicular
networks have not been fully developed yet, they could be used in a near
future as a means to provide a number of interesting applications and
services that need the exchange of data among vehicles and other data
sources.

For example, we can consider the collection of information within an
interesting area in a city using a spatial crowdsourcing schema that takes
advantage of the network formed by the vehicles, as well as the interests of the people that travel aboard them. In this paper, we present a
preliminary spatial crowdsourcing approach that uses the technology of
mobile agents to accomplish the collection and querying of data in such a
scenario, supported by realistic simulations that prove that the proposal
is promising.

1 Introduction

In recent years, the development of information and telecommunication technologies have reached the automotive industry, giving the vehicles the ability of
sending and receiving data, and thus they can form what is known as a vehicular
ad hoc network or VANET [6]. In this type of networks, short-range communications are used (because of its simplicity and low cost) to exchange data among
cars and other actors (e.g., pedestrians or the road-side infrastructure).

The average modern vehicle has a number of on-board sensors that provide
the data needed for a proper and safe operation, but it could also be equipped
with other types of sensors to get information of the surroundings (e.g., temperature, humidity, pollution gases, etc.). In this way, they could become mobile
sensing platforms that could take measures of parameters in any place of the
city or area where they drive. The amount of data collected in this way can be
quite high, and interesting *knowledge* can be extracted if the data are properly
analyzed. However, some difficulties arise: What happens if we are interested in

© Springer International Publishing Switzerland 2016
M. Ivanović et al. (Eds.): ADBIS 2016, CCIS 637, pp. 88–95, 2016.
DOI: 10.1007/978-3-319-44066-8_10

a certain area and the amount of available data is too low? How can we retrieve such data using the limited short-range communications used in the VANET?

In this paper, we present a system that obtains information stored on vehicles equipped with sensors while they are moving along the roads of a city or region. To accomplish this, we use mobile agents that use spatial crowdsourcing to ease the process of locating and retrieving the required data.

The rest of this paper is structured as follows. In Sect. 2, we describe the basics of intelligent vehicles, vehicular networks, and their applications. In Sect. 3, we describe a system for monitoring and querying data using a VANET. Section 4 describes the use of crowdsourcing techniques to improve the monitoring process. In Sect. 5, we perform an experiment to evaluate the feasibility and interest of the proposal. In Sect. 6, we compare some related works with ours. Finally, in Sect. 7, we summarize our conclusions.

2 Intelligent Vehicles and Vehicular Networks

In the last years, cars and other vehicles have progressively included more and more devices, similar to conventional computers, to make the driving process more comfortable and easy. For example, GPS navigators and multimedia centers have become very popular, but despite the underlying technology is quite mature other applications are very scarce yet.

The most popular applications in the Internet are social-like (e.g., Facebook, Twitter, LinkedIn) as well as collaborative or crowdsourcing applications. In these applications a number of individual users contribute by giving a relatively small amount of resources by means of a network connection (the Internet) and, by combining these contributions, a certain objective or problem is solved or achieved. For example, in OpenStreetMaps[1] they share knowledge about the streets of the cities to build free-to-use maps, and in the Folding@Home project[2] they share computing power to perform biochemical-related simulations.

Regarding vehicular-related applications, we have a number of similar elements to those of the crowdsourcing scenario, i.e., *users* that may be willing to share, using a *network*, some *resources* related to driving vehicles (e.g., routes, traffic status, etc.). However, this can be problematic, since the only way of accessing directly to the Internet is the use of 3G/4G connections, which have an economic cost and do not cover all areas (e.g., tunnels and undergrounds).

To solve this problem, Vehicular Ad Hoc Networks (or VANETs) [6] have been proposed. In these networks, the vehicles act as nodes which send and receive data from other nearby vehicles. The link between those nodes is established directly in a peer-to-peer way using any of the available short-range wireless communication technologies (e.g., Wifi, WiMAX, or the standard WAVE [3]).

When a certain amount of data are transferred using a VANET, they must be sent from one node (vehicle) to another, and this process is repeated until the data reach their destination. Since the vehicles are constantly moving these

---

[1] http://openstreetmaps.org.
[2] http://folding.stanford.edu.

connections may last only a few seconds and therefore the network topology is constantly changing, making it very difficult to route the data.

3   Query Processing in VANETs

One proposed application for VANETs is the monitoring of certain parameters in an area by the moving cars present in it [10], and the possibility of submitting queries about information related to that area that are processed using the most updated data present in the VANET.

In this scenario, a percentage of vehicles are assumed to carry a portable computer which collects data about their surroundings, that may be of interest to other users (e.g., information about available parking spaces, prices at gas stations, etc.). It would also be very interesting if the vehicles carried different types of sensors to collect information in different places where the vehicle may travel to (e.g., temperature, noise, pollution gases, etc.).

The acquired data are stored locally and can grow quickly if both the number of sensors and the sampling rate are high. It may not be possible to transfer all that data to a central place, given the limitations of the network connections and also that only a small amount of the collected data would likely be valuable.

For these reasons, we proposed a way of processing the data locally on the vehicles by using mobile agents. A mobile agent [5] is software that can move from one computer to another using a network connection. For this to be possible, it is necessary that all the involved computers execute a middleware called the *mobile agent platform* [9], which manages the movements of the agents and provides them a number of additional services.

The whole query processing would consist of the following four steps:

1. First, the user creates a query about some information of his/her interest related to an area (e.g., the pollution levels in the city center), and immediately a mobile agent starts its execution with the defined query parameters.
2. In the second step, the mobile agent travels towards the interest area using the vehicles in the VANET to *hop* from one vehicle to another, until it reaches the destination. Every time the mobile agent arrives at a vehicle, the next hop may not occur immediately, since the number of potential vehicles to hop to might be insufficient. In such cases, the mobile agent will stay in the same vehicle waiting for a more suitable one, or even use it as a *taxi* that will physically carry the mobile agent closer to its destination area.
3. Once there, the agent hops among the vehicles within the area and, upon its arrival to each of them, processes the locally-stored data on that vehicle to try to find an answer (total or partial) to the query initially created.
4. Finally, in the last step, the mobile agent returns the result of the query to its originator using the same procedure of hopping among vehicles.

One advantage of using mobile agents is that, instead of sending only data, they can carry along with them the logics or algorithms needed to transfer and process that data: Since the vehicles are constantly moving, it is not possible to use routing tables to reach the destination. Therefore, it is necessary to evaluate continuously the appropriate next hop according to the traffic conditions and other factors. Once the data is reached, it may be processed differently according to its nature (e.g., concentration of pollution gases, gas prices, etc.). The ability of mobile agents to carry such algorithms (that can additionally be changed or enhanced at any moment) makes them a very flexible and functional option.

We believe that this is an interesting application for VANETs, that makes it possible to extract valuable information using the vehicles as moving sensing platforms, which is more efficient and dynamic than the traditional solution based on measurement stations deployed at fixed locations. Besides, mobile agents are very adequate since their flexibility and autonomy provide advantages in a fast-changing scenario such as a VANET. To reinforce this idea, we performed some initial experiments to evaluate different features [10], which show that this proposal is promising.

4   Use of Spatial Crowdsourcing for Query Processing

The previously-described query solving process uses the features of mobile agents to solve queries in VANETs, but it also has some drawbacks: If the traffic density is too low, the mobile agent may not find a path to reach the interest area in a reasonable time. Besides, if the carrying vehicle leaves the area, the mobile agent could get *lost*, interrupting the query solving process.

To overcome these problems the mobile agent could be *cloned* into several copies of itself to maximize the probability of reaching the destination following different routes, and additionally a *timeout* can be established so that if the agent does not return before the time expires, it is supposed to be lost and it may be launched again or any other further actions can be taken.

Another strategy for enhancing the query solving process is the use of spatial crowdsourcing techniques. For example, if the mobile agent finds it difficult to reach the interest area, it can get help from other drivers who would be compensated using some kind of *virtual money* or other benefits. This would take place in the second step of the process followed to solve the query (see Sect. 3), where the mobile agent would try to get help from the drivers that previously had stated that they were willing to collaborate. For example, some drivers could detour from their route to carry physically the mobile agent near its destination.

Another crowdsourcing strategy not involving changes in the expected routes could be that certain drivers (e.g., commuters) could make available the route they will follow (the whole route or only a portion), to help the mobile agent to decide if that vehicle is worth to hop to or not.

5   Preliminary Experimental Evaluation

In this section, we present some preliminary experiments that we have performed to test the feasibility of enhancing the query solving process in VANETs using mobile agents with spatial crowdsourcing abilities. For evaluation, we have used the MAVSIM simulator [11], that allows the simulation of both mobile agents and traffic using real road maps extracted from OpenStreetMaps. Among its various functionalities, we can highlight the simulation of wireless signal blocking by buildings, as this is more realistic.

5.1 Experimental Setup

These are the parameters of the simulations considered in the experiments that we have carried out:

- The *map scenario* is a portion of 16 Km^2 belonging to the city of Madrid (Spain), with a balanced ratio of long straight road segments and short and/or curved segments, that neither favors nor harms the wireless signal propagation (see Fig. 1).
- In this experiment, the mobile agent applies a greedy strategy trying to get closer to the destination as soon as possible.
- Once the mobile agent reaches the destination area (which has a surface of 0.25 Km^2), we assume that it should collect five *results* from the vehicles which contain information *relevant* to the query, being the ratio of such vehicles only the 50 %.
- When the mobile agent finds a suitable vehicle, it takes five seconds to process the stored data and extract only the portions needed to solve the query.
- To avoid that the data collection takes too long, a timeout of 60 s is set. If it is reached, the mobile agent will return to the query originator with the data collected so far, that can possibly be less than desired.
- When the mobile agent hops from one vehicle to another, it takes *one second* to transfer. A large number of factors can affect the effective speed (e.g., interferences, obstacles, the overhead due to the serialization of the agent, code and data, etc.). Therefore, we do not assume a best-case scenario and establish this value of one second, which we consider reasonable according to field experiments that we have performed with real smartphones and tablets.

With this setup, we repeat every simulation 50 times with different random starting positions for the vehicles, and compute the average of the results. The experiments performed are described in the rest of the section.

5.2 Showing the Interest of Spatial Crowdsourcing

In this experiment, we evaluate the potential benefits of spatial crowdsourcing. We simulate the collection of environmental data within an area that is crossed by only 5 % of the vehicles. In such scenario, a spatial crowdsourcing approach, rather than just relying on the usual flow of vehicles, may be required.

To encourage the collaboration from other vehicles, the mobile agent pays *virtual money* to vehicles in exchange of being transported closer to the target area (the *negotiation* to alter the driver's current trajectory is supposed to take place automatically). We vary the ratio of collaborating drivers from 0 % (i.e.,

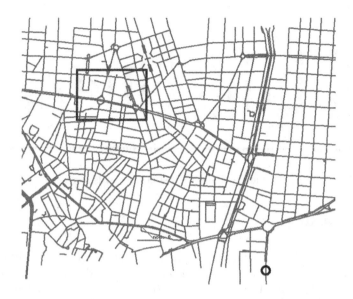

Fig. 1. Scenario map used in the simulation. The rectangle shows the designated interest area, and the small circle the starting point where the query is launched

the mobile agent does not exploit spatial crowdsourcing) to 100 % (i.e., every
driver is willing to collaborate).

The amount of money spent by the agent is proportional to the effort performed by the vehicle. Specifically, the mobile agent asks the driver to be transported to at most 500 meters from the center of the interest area, and the
amount paid will be proportional to the total distance that the agent has been
transported.

In Fig. 2(a), the total time required to collect the data is shown, which
involves following all the steps of the monitoring process described in Sect. 3. As
the amount of participating vehicles increases, less time is required to complete
the process, as spatial crowdsourcing can help in this case to reach the target
area faster than by just hopping among vehicles that are not likely to travel
there. According to the results, about 40 % of participating vehicles is enough
and a higher percentage of drivers willing to participate does not contribute to
decreasing significantly the time spent in the process.

In Fig. 2(b), the number of hops performed by the mobile agent to complete
the process is shown. Since every time the agent hops it must be transferred
using a wireless connection, these values give an idea of the bandwidth used in
the process. When the number of collaborating vehicles is low, the agent must
constantly hop among the vehicles to find a route to reach the interest area.
When the number of collaborating vehicles is higher, it is more likely that the
agent will be transported near the interest area, decreasing the number of hops
needed and the bandwidth. However, it should be noted that hopping through
the wireless network is quicker than being physically transported by a vehicle.

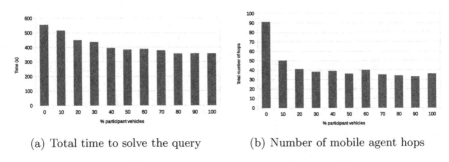

(a) Total time to solve the query   (b) Number of mobile agent hops

Fig. 2. Query solving with spatial crowdsourcing

6   Related Work

The topic of spatial crowdsourcing has attracted significant research attention (e.g., see [2,8]). Besides, a recent work analyzes crowdsourcing in Intelligent Transportation Systems (ITS) [12], but it does not focus on vehicular networks nor spatial crowdsourcing. Up to the authors' knowledge this is the first proposal that applies the idea of spatial crowdsourcing to vehicular networks.

In [4], the authors assume the existence of a vehicular network in which the vehicles carry sensors, and they pose the problem of determining how much information should be received from the sensors and where they should be located to perform an optimum monitoring process. However, the data are sent to a server using 2G/3G connections, which has some disadvantages, such as the economic cost.

A working prototype is presented in [7] that consists of a monitoring VANET, where the vehicles are equipped with sensors. The collected data are sent opportunistically to a central server using Wifi access points. By contrast, in our approach the data stored in the vehicles are processed distributively by mobile agents searching for the most relevant data.

Finally, in [1], the authors present a sensing engine called MoST, that can be used as a framework for developing location-aware applications for Android mobile phones. This work it is not related specifically to VANETs, but it illustrates the interest of sensing and crowdsourcing applications.

7   Conclusions and Future Work

In this paper, we have presented the basis of a method to use crowdsourcing techniques to enhance the processing of queries in a VANET using mobile agents. When they are performing their task, they have the possibility of asking for help to other users that can collaborate, for example, by modifying their vehicle's route to transport the agent nearer its destination. We have performed some tests and the results obtained show that the use of this technique could increase the performance of the process, allowing the agent to reach the interest area faster, with a more efficient use of the wireless bandwidth than when these techniques are not used.

As future work, we will analyze the use of other spatial crowdsourcing techniques and query processing strategies and their evaluation in a wide variety of scenarios. The promising results that we have obtained with this first approach encourages us to continue this research line and explore these and other possibilities.

Acknowledgments. The authors acknowledge the support of the CICYT project TIN2013-46238-C4-4-R and DGA-FSE.

References

1. Cardone, G., Cirri, A., Corradi, A., Foschini, L., Ianniello, R., Montanari, R.: Crowdsensing in urban areas for city-scale mass gathering management: Geofencing and activity recognition. Sens. J. IEEE **14**(12), 4185–4195 (2014)
2. Chen, Z., Fu, R., Zhao, Z., Liu, Z., Xia, L., Chen, L., Cheng, P., Cao, C.C., Tong, Y., Zhang, C.J.: gMission: A general spatial crowdsourcing platform. Proc. VLDB Endowment **7**(13), 1629–1632 (2014)
3. Jiang, D., Delgrossi, L.: IEEE 802.11p: Towards an international standard for wireless access in vehicular environments. In: IEEE Vehicular Technology Conference (VTC Spring), p. 2036–2040 (2008)
4. Liu, L., Wei, W., Zhao, D., Ma, H.: Urban resolution: New metric for measuring the quality of urban sensing. IEEE Trans. Mobile Comput. **14**(12), 2560–2575 (2015)
5. Milojicic, D., Douglis, F., Wheeler, R.: Mobility: processes, computers, and agents. ACM, New York (1999)
6. Olariu, S., Weigle, M.C., Networks, V.: From Theory to Practice, 1st edn. Chapman and Hall/CRC, Boca Raton (2009)
7. Santos, P.M., Calcada, T., Guimarães, D., Condeixa, T., Sargento, S., Aguiar, A., Barros, J.A.: Demo: Platform for collecting data from urban sensors using vehicular networking. In: Proceedings of the 21st Annual International Conference on Mobile Computing and Networking, (MobiCom), pp. 167–169. ACM (2015)
8. To, H., Ghinita, G., Shahabi, C.: A framework for protecting worker location privacy in spatial crowdsourcing. Proc. VLDB Endowment **7**(10), 919–930 (2014)
9. Trillo, R., Ilarri, S., Mena, E.: Comparison and performance evaluation of mobile agent platforms. In: The Third International Conference on Autonomic and Autonomous Systems (ICAS), pp. 41–46. IEEE Computer Society (2007)
10. Urra, O., Ilarri, S.: Using mobile agents in vehicular networks for data processing. In: 14th International Conference on Mobile Data Management (MDM), vol. 2, pp. 11–14. IEEE (2013)
11. Urra, O., Ilarri, S.: 10 MAVSIM: Testing VANET Applications Based on Mobile Agents, pp. 199–224. CRC Taylor and Francis Group, 2016. Print ISBN: 978-1-4987-2191-2, eBook ISBN: 978-1-4987-2192-9. doi:10.1201/b19351-14
12. Wang, X., Zheng, X., Zhang, Q., Wang, T., Shen, D.: Crowdsourcing in ITS: The state of the work and the networking. IEEE Trans. Intell. Trans. Syst. **17**(6), 1596–1605 (2016)

cIPT: Shift of Image Processing Technologies to Column-Oriented Databases

Tobias Vinçon[1,2(✉)], Ilia Petrov[2,3], and Christian Thies[3]

[1] Hewlett Packard Enterprise, Böblingen, Germany
tobias.vincon@hpe.com
[2] Data Management Lab, Reutlingen University, Reutlingen, Germany
[3] Reutlingen University, Reutlingen, Germany
{tobias.vincon,ilia.petrov,christian.thies}@reutlingen-university.de

Abstract. The amount of image data has been rising exponentially over the last decades due to numerous trends like social networks, smartphones, automotive, biology, medicine and robotics. Traditionally, file systems are used as storage. Although they are easy to use and can handle large data volumes, they are suboptimal for efficient sequential image processing due to the limitation of data organisation on single images. Database systems and especially column-stores support more stuctured storage and access methods on the raw data level for entiere series.

In this paper we propose definitions of various layouts for an efficient storage of raw image data and metadata in a column store. These schemes are designed to improve the runtime behaviour of image processing operations. We present a tool called column-store Image Processing Toolbox (cIPT) allowing to easily combine the data layouts and operations for different image processing scenarios.

The experimental evaluation of a classification task on a real world image dataset indicates a performance increase of up to 15x on a column store compared to a traditional row-store (PostgreSQL) while the space consumption is reduced 7x. With these results cIPT provides the basis for a future mature database feature.

1 Introduction

Industries like automotive, biology, medicine and robotics produce more media data than ever nowadays. Trends in mobile devices and social media platforms additionally increase the amount of image data dramatically. Furthermore, the need to process this image data has risen up to real time applications in the last years, driven mainly by big data analyses [3].

Especially, the uniform analysis of image sequences is very wide-spread across emerging image processing applications for instance content based image retrieval, pedestrian tracking in intelligent cars or cell recognition in series of micrographs. Most traditional image processing tools use file systems, which are suboptimal persistence layers because algorithms cannot operate sequentially on specific data sections over all images without costly loading of every single image

© Springer International Publishing Switzerland 2016
M. Ivanović et al. (Eds.): ADBIS 2016, CCIS 637, pp. 96–103, 2016.
DOI: 10.1007/978-3-319-44066-8_11

entirely into memory. DBMSs can make this data access more transparent with respect to local structures and thus offer the possibility to execute operations more efficient and closer to the data. However, traditional row-stores cannot efficiently handle the high volume data streams resulting from image sequences. Modern column-stores are well-suited for the needs of big data analysis [1].

This paper proposes the column-store Image Processing Toolbox (cIPT) by providing the ability to process image sequences on a column-oriented DBMS. cIPT allows to use different data layouts for the image data extracted from an input sequence. These data layouts leverage the characteristics of a column-store with a set of basic image processing operations. This approach is illustrated by implementing an image analysis pipeline for image classification which is the underlying principle of the applications mentioned above.

Data comes from the Cape2Cape project [5] (a cooperation between the German car manufacturer Volkswagen and Hewlett Packard) where a customized production car drove from the north to the south cape in under 9 days. The vehicle was equipped with a front camera configured to take high definition images every 10 s. The goal is to analyse the images in near-time, to extract information about the weather condition at the respective location and to correlate them with global weather forecasts. The result is: high volume image data, the need for optimal persistence and acceleration of image processing operations. The images have to be classified with respect to typical weather conditions.

The required image analysis pipeline is adapted to characteristics of the column-oriented database where the first version of cIPT includes: an image load utility, data layout converters, neighbourhood access functionality, extraction and comparison of histogram data as well as support for different distance metrics.

In this sense, cIPT provides a basic toolkit functionality for building image processing applications based on columnwise data access. Further contributions of this paper are the implementation of cIPT within a commercial column-store. The performance is evaluated on the real Cape2Cape image data set, which amounts to 183GB and comprises to approx. 70000 images. The experimental evaluation compares cIPT on a column-store against a traditional row-store (PostgreSQL).

The rest of this paper is organized as follows. The next Section presents the related work. The architecture of cIPT and the different data layouts as well as the definition of the operations is described in Sect. 3. The experimental evaluation is presented in Sect. 4. We summarise our results and conclusions in Sect. 5.

2   Related Work

Since images are not only analysed individually but are rather considered as a source of observation data, their processing finds increasing application in many industrial areas [2,7].

The shift of the persistence layer of media storage applications from a conventional file storage to a relational database began in the early 90 s with IBM's

QBIC project [9]. It provides an automatic feature extraction of loaded images and a similarity search on their basis. The runtime of utilized image processing algorithms is improved significantly by placing them in the DBMS. By storing only the extracted features like histograms, QBIC applications are able to search images within large collections but cannot efficiently perform repeating feature calculations with varying settings.

Recent research in the area of DBMS proves that column-oriented databases outperform traditional row-stores when handling data-intensive workloads as demonstrated by C-Store (nowadays Vertica) [8,11]. Efficiently utilised techniques like data compression, encoding, late materialisation, main-memory processing and support for parallelisation characterise the architecture of column-stores. Modern column-stores are able to handle mixed loads and update-intensive operations while performing complex analytical queries by applying various optimizations (e.g. the distinction between read- and write-optimised stores in Vertica or delta stores in other stores, etc.). MonetDB is a widely used open-source column-store. [6] illustrates that processing approximately 4 TB of image data within the Sloan Sky Server project, MonetDB has lower response times than a commercial row-store (MS SQL Server) for almost all queries.

Due to their characteristics, column-stores are to process high volume image sequence data. Under cIPT we purpose an approach for a basic toolset of data layouts and operations as basis for image processing and detection algorithms. cIPT is optimised for column-oriented DBMSs and it has the potential for a major database feature in the future.

3   Architecture of cIPT

The column-store Image Processing Toolbox (cIPT) is a generic set of data layouts and algorithms for the purpose of image processing on column-oriented databases. Although existing functionality focuses on the comparison of images by calculating similarities in their colour values, further more general research questions on image processing on column stores can be answered. Moreover, cIPT has a modular and extensible design.

Features e.g. colour histograms, are derived from raw pixel data for instance for similarity search [2]. This toolbox currently uses RGB colour values and their average greyscale. The distribution of these values over all images' pixels is represented by histograms. If the application needs to extract a specific part of an image, cIPT will be able to filter the raw data very fast.

Usually, image processing algorithms behave in a common manner, based on a general workflow. The following enumeration sketches its steps, data interaction and its utilisation within the cIPT.

1. **Load** The load operation of cIPT extracts the RGB colour values of each pixel using an integrated C++ image processing API. If necessary, further information like the greyscale is calculated on the fly. The entire data is represented as vectors in the Tables *(rgb)*, *(rgbgrey)* or *(grey)*. This complete process is covered in a special module called User Defined Load (UDL).

2. **Convert** *(optional)* The conversion of image data from one relation to another is an optional operation that requires further calculations in special cases. Several conversion operations are designed within cIPT.

3. **Filter** *(optional)* Filtering is yet another optional operation within cIPT. It takes an image data table as input and extracts image data within certain shapes (e.g. rectangles, circles, etc.). The result has the input format, allowing pipelining.

4. **Calculate Histogram** The calculation of histograms is influenced by several execution parameters. Using minimum, maximum and the number of bins the detail degree is adjustable. These histograms are persisted and reused in further processing steps.

5. **Compare** The similarity of histograms can be calculated in several ways. For instance, the distance metric can weight big divergences much more than small ones or conversely. There are several similarity metrics available in cIPT (manhatten, euclidean, etc.).

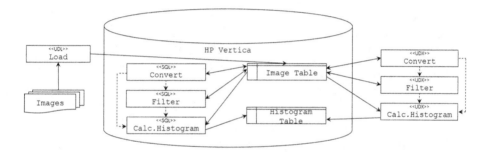

Fig. 1. General overview of relations and operations within the cIPT

Possible interactions within cIPT are sketched in Fig. 1. cIPT includes **Image Tables** as raw data source and **Histogram Tables** as storage for calculated features. Both types of relations are discussed in detail in Sect. 3.1.

3.1   Data Schema Layout

The cIPT schema comprises two types of relations. Firstly, the image table stores images raw data, more precisely the colour representations *rgb*, *rgbgrey* and *grey*. Secondly, the histogram table stores the calculated histogram data including their properties. Tuples of both relations can be arranged in a horizontal, vertical or in partitioned manner causing performance differences. Although a partitioned layout might be efficient in some cases where queries only access data of one partitioned table as shown in [10], the cIPT usually accesses either the complete data of an image or an undefined portion of it. Therefore, partitioned layouts do not suit cIPT operations.

In addition to the payload, each table reserves space for an unique identifier and further metadata as shown exemplarily for the vertical *rgbgrey* layout in Table 1[1]. Unfortunately, the commercial column store we used for the performance evaluation is limited to a maximum of 1600 columns which is the reason an unpartitioned-horizontal layout is partially possible. If implemented, one image is represented by a single tuple comprising columns for each pixel. Because of the limitation, only images with resolution of smaller than 40×60 pixels can be stored which is obviously not suited for modern applications. However, histogram data can be arranged both vertically and horizontally. The size of cIPT's histograms depends primarily on the colour depth. 256 containers are sufficient to store images with 8 bit colour depth. In addition, characteristics like the colour channel, the number of bins, minimum and maximum colour values are stored. An example of the vertical arrangement is shown in Table 2. In the case of a horizontal arrangement, bins are no longer represented as a set of tuples but rather as different columns (cf. Table 3). One obvious advantage of the horizontal layout over the vertical is the reduced redundancy.

Table 1. Schema of RGB_Grey

imageid	x	y	red	green	blue	grey
...	...	...	...	...	...	...

Table 2. Vertical histogram schema

id	chan.	bin#	min	max	bin	freq
...	...	256	...	...	0	...
...	...	256	...	...	...	...
...	...	256	...	...	255	...

Table 3. Horizontal histogram schema

id	chan.	bin#	min	max	bin_0	...	bin_255
...	...	...	...	...	...	...	...

Table 4. Operation implementations

Operation	Column Store	pgSQL
Load (rgb)	UDF	-
Load (rgbgrey)	UDF	-
Load (grey)	UDF	-
Convert (rgb to rgbgrey)	UDF/SQL	-
Convert (rgb to grey)	UDF/SQL	-
Filter rectangle	SQL	SQL
Filter circle	SQL	SQL
Calc. histogram_v	UDF/SQL	SQL
Calc. histogram_h	UDF/SQL	SQL
Average histogram_v	SQL	SQL
Average histogram_h	SQL	SQL
Euclidean histogram_v	SQL	SQL
Euclidean histogram_h	SQL	SQL
Manhattan histogram_v	SQL	SQL
Manhattan histogram_h	SQL	SQL

3.2 Operation Definition

Processing images requires a set of operations as presented in the basic cIPT architecture. In an application, histograms might be compared to each other or to a previously created average histogram. Operations for these kinds of workflows are provided within the cIPT. They are implemented in either C++ and exposed to DBMS as User Defined Functions (UDF) or as SQL statements. Table 4 presents the existing functionality of cIPT for the commercial column-store and PostgreSQL. The parameters of UDFs and SQL implementations are equal. Their performance is evaluated in the following Sect. 4.

4   Experimental Evaluation

The evaluation of the cIPT is based on multiple defined test scenarios. Each of them is executed on the same system with same *original* dataset of 100 images

---

[1] Similar layouts exist for the *rgb* and *grey* relation.

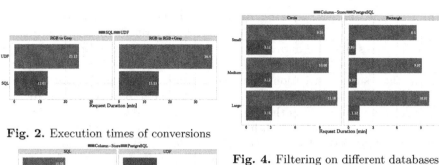

Fig. 2. Execution times of conversions

Fig. 3. Execution times of different oriented histogram calculations

Fig. 4. Filtering on different databases

Fig. 5. Space consumption of histograms

from the Cape2Cape challenge. A few of the most interesting results are listed below.

Is the compression factor of a column store equally effective as the one of a row store? To test this research question the *original* data set is loaded into the column store and PostgreSQL database. After the execution the space consumption of all relations are reported. Comparing these measurements clearly shows that the space consumption in the row store is dramatically worse. This is because PostgreSQL requires, even though no index is defined, space for tuple and page header. In addition it reserves about 10 % to 20 % of each page's available space to increase the runtime of updates at a later time.

Is the calculation of missing information more reasonable during the initial load or with a further processing step, utilized either with an UDF or via SQL? Additional information can be calculated on the fly throughout the load- or a further processing-step. This experiment measures representative the execution times of UDFs and SQLs calculating the missing grey layer and compares these with a full load of the rgbgrey representation. Figure 2 demonstrates that a SQL implementation is with a factor of 0,42 to 0,51 dramatically faster than the UDFs explainable by the closer operation of SQL within the database's layout. However, UDFs ability to express complex custom implementation is not negligible. Regarding the postponed calculation of information at all, the clear suggestion to compute as much information as possible during the load can be provided. Every further processing step has to read the entire data. Yet, the load operation has this data present as well during its execution and avoids unnecessary reads.

Which is the fastest filtering option on both databases? Different filter operations are investigated within this test scenario. Therefore the *original* dataset provides the base data on which rectangles and circles are extracted in various sizes by SQL implementations on the column and the row store. Figure 4 presents the resulting execution times of every filter operation. Obviously recognisable is the significant difference between the both database types. The column store takes only up to 10 % of the time the row-oriented database consumes. This clear advantage might scale with increasing data. Moreover, filtering less complex shapes like rectangles improves the run time of these operations as well. The correlation between execution time and filtered size is less a influence factor and can be broken down to increased data transfer.

Which database extracts features more efficiently, either with a UDF or SQL? This test scenario measures the duration of cIPT's feature extraction operation on both database types. Therefore a UDF and SQL implementation creates vertical and horizontal histograms of the *original* dataset's images. In Fig. 3 the constant duration of UDFs with 4–9 minutes is recognisable. The results of the SQL implementation are very wide spread. After analysing the biggest outlier with more than 300 min reveals that the multiple DECODE statements are highly inefficient. Using a UDF to transpose the result of a vertical histogram operation as a work-around decreases the execution time dramatically. The remaining measurements conclude that the column-store performs better on creating histograms both with SQL and UDFs.

What is the space consumption with different histogram orientation types on a row and a column-store? The last test scenario analysis the space consumption of histograms with different orientations and image areas as input. Figure 5 depicts the measurements for every type on both database types. The correlation between the input size and space consumption can be explained by the implementation. Since the probability that a bin becomes assigned decreases with the shrinking image area, the amount of unassigned bins increases. These bins can either be compressed very efficiently as their value is 0 or there are non-existing e.g. vertical layout. While the commercial column-store profits by increasing the compression rate, PostgreSQL with its fixed size per cell is limited to the 120 KB in space consumption. The most efficient combination is as expected the horizontal layout on the column-store because the layout adapts the characteristics of the column-store and can compress the data to tiny 33–37 KB.

5   Conclusion

The development of the cIPT yield a first prototype of an image processing feature for column oriented databases. Evaluating the included different-oriented data layouts on PostgreSQL as a row-store and a commercial column-store reveals the advantages and disadvantages of each. Furthermore, the much better compression of the column-store is clearly visible in both, the loaded image data and calculated features.

Furthermore, experiments with the cIPT's operations analyse the differences between implantations of proprietary UDFs and queries using SQL. UDFs are constructed for very flexible use cases and complex calculation which could not be put into practise via SQL. Especially the loading process of the cIPT is realised with a special UDL, implementing several open-source C++ libraries for image processing. However, queries in SQL are significantly faster since they operate much closer on the database system and are efficiently restructured by the database's build-in query optimizer.

Using cIPT for a real world scenario, classifying recorded images during a road trip from the north to the south cape by a mounted camera shows one possible application.

References

1. Abadi et al.: Column-oriented database systems. VLDB, August 2009
2. Datta, R., et al.: Ideas, influences, and trends of the new age. ACM Comput. Surv. **40**(2), 5:1–5:60 (2008)
3. Deligiannidis, L., Arabnia, H.: Emerging Trends in Image Processing, Computer Vision and Pattern Recognition. Emerging Trends in Computer Science and Applied Computing. Elsevier Science (2014)
4. Deselaers, T.: Features of image retrieval, December 2003
5. HP. Cape2cape. http://www8.hp.com/uk/en/campaigns/cape2cape/overview.html
6. Ivanova, M., et al.: Monetdb/sql meets skyserver: the challenges of a scientific database. In: Proceedings of the SSBDM (2007)
7. Johansson, B.: A survey on: Contents based search in image databases. Survey, Department of Electrical Engineering, Linköping University 08 (2000)
8. Lamb, A., et al.: The vertica analytic database: C-store 7 years later. Proc. VLDB Endow. **5**(12), 1790–1801 (2012)
9. Niblack et al.: querying images by content, using color, texture, and shape (1993)
10. Sidirourgos, L., et al.: Column-store support for rdf data management: Not all swans are white. Proc. VLDB Endow. **1**(2), 1553–1563 (2008)
11. Stonebraker, M., et al.: C-store: A column-oriented dbms. In: Proceedings of the VLDB (2005)

Influence of Parallelism Property of Streaming Engines on Their Performance

Nigel Franciscus[1], Zoran Milosevic[2], and Bela Stantic[1(✉)]

[1] Institute for Integrated and Intelligent Systems,
Griffith University, Queensland, Australia
B.Stantic@griffith.edu.au
[2] Deontik, Brisbane, Australia

Abstract. Recent developments in Big Data are increasingly focusing on supporting computations in higher data velocity environments, including processing of continuous data streams in support of the discovery of valuable insights in real-time. In this work we investigate performance of streaming engines, specifically we address a problem of identifying optimal parameters that may affect the *throughput* (messages processed/second) and the *latency* (time to process a message). These parameters are also function of the parallelism property, i.e. a number of additional parallel tasks (threads) available to support parallel computation. In experimental evaluation we identify optimal cluster performance by balancing the degree of parallelism with number of nodes, which yield maximum throughput with minimum latency.

1 Introduction

In Big Data environment, there are two types of data processing engines dedicated for different purposes, namely batch and streaming engines. Batch processing is concerned with the handling of massive volume of data while streaming is concerned with processing data of high velocity. Several platforms and tools have been developed to support big data environments, of which the most widely known is Hadoop[1] which utilizes MapReduce batch processing. The Hadoop is however not adequate for real time streaming requirements, as for example is needed for stock market pattern monitoring, ad-serving [5] or real-time personalised recommendation[2]. Total time T to process a message is

$$T = B + M$$

where B is the time to collect data in the Hadoop input buffer and M as the time to process the data through MapReduce. It is obvious that some applications require near real-time analysis and this type of analysis can be supported by streaming engines. Two most popular Streaming Processing Engines (SPEs) are

---

[1] Apache Hadoop, http://hadoop.apache.org/.
[2] Spotify Labs, https://labs.spotify.com/2015/01/05/how-spotify-scales-apache-storm/.

© Springer International Publishing Switzerland 2016
M. Ivanović et al. (Eds.): ADBIS 2016, CCIS 637, pp. 104–111, 2016.
DOI: 10.1007/978-3-319-44066-8_12

Storm and Spark Streaming which are fault tolerant and guarantee message delivery.

Due to simultaneous threads execution, it is expected that degree of parallelism will improve the throughput. However, it has been mentioned in literature that when the degree of parallelism reaches certain threshold there may be a latency penalty which will impact overall performance [7]. In order to identify the optimal configuration and performance of SPEs with varying degree of parallelism, we have carried out a series of experiments across the clusters in Storm and Spark and looked closely in throughput and latency performance parameters. In experimental evaluation, which is based on counting words extracted from sentences submitted to the SPEs in continuous stream, we varied different *degrees of parallelism* and *number of nodes* used. In addition, we investigated a possible relation between network bandwidth and threshold that each engine can support. We have not considered the impact of the CPU usage and memory consumption.

2 Related Work

To the best of our knowledge, there is no existing work focusing on the Streaming Processing Engines regarding the degree of parallelism. However, some of the benchmarking have been done in term of performance.

Work on Discretized Stream provided a performance analysis of Spark and Storm throughput [9,10]. Bell-Labs explored Storm inner core to get Storm performance optimisation by looking at parameter configuration [1,7]. The inner core that has been tested includes number of parallelism hint, network usage, CPU usage and horizontal scalability. Bell - Labs have found that the configuration is critical and they are planning to build an engine which can determine the best configuration for every job automatically. The key attribute to consider in the implementation of the engine is the horizontal scalability configuration.

Our work is built on this specific criterion of *parallelism degree* which exploits the benefit of distributed systems by increasing the number of parallel threads and distribute them reliably across the nodes. However, it has been shown that the effect of spout-bolt parallelization has reach level of threshold with 24° of parallelism. Our work proved that we can overcome the level of threshold by investigating network bandwidth. Note that threshold is also dependent on environment hardware.

Analysis presented in [8] describes the use of Storm within Twitter including Storm architecture and methods. An investigation of comparison between Apache Storm and Apache Flink (Formerly known as Nephele) was presented in [3,4] in relation to latency constraint and throughput. Since latency is a non-trivial property in streaming context, this work has studied on how the system accommodates such latency restriction while producing maximum throughput. While study conducted in [2,6] provided an architectural comparison of Spark and Storm, it concluded that Storm performs well in sub-second *latency* and no data loss while Spark performs better where *throughput* and stateful processing is required.

3   Influence of Parallelism Property

We have investigated the dependency of throughput and latency measures in streaming engines considered from (i) the degree of parallelism parameters and ii) number of worker nodes as variables. We performed the analysis on the individual engine basis which subsequently led us to establish some comparative results of one-at-the-time and micro-batch processing. The purpose of these experiments was to better understand the interplay of various parameters that are configurable in both engines and various choices that can be adopted to support performance tuning.

Our experiment was conducted on a cluster of 20 nodes interconnected via Giga Ethernet. Each node has quad-core Intel(R) Core(TM) i5-2400 CPU @ 3.10 GHz with 4 GB RAM and 500 GB SATA HDDs. The nodes run Scientific Linux release 6.3 and are connected using standard 100 Mb/s Ethernet and 1 Gb/s Ethernet. For the experiment, datasets are handled by Apache Kafka, where messages are injected into Kafka to build up the queue. Our method to collect the statistics of *throughput* and *latency* is during the peak time of the message being processed. Thus, it will allow engine to apply the latency constraint while trying to achieve maximum throughput.

For each collection, we calculate the mean of the average throughput-latency within the interval of biggest output with the average of 5 min after job submission. We use both Storm and Spark metrics to collect statistics for both throughput and latency. In this experiment, we use Apache Storm 0.9.3, Apache Spark 1.4.0 and Apache Kafka 0.8.2.1. Both Apache Storm and Spark are given 4 slots (executors) allocation for each node.

As mentioned before, we use word count job as the measurement benchmark for testing Storm and Spark Streaming capabilities. This job imposes high processing demand on CPU resource. Each experiment is tested within the same cluster and same configuration for *degree of parallelism* parameter to facilitate comparison. Both engines have the degree of parallelism parameter that defines replication of receiver and processor to handle multiple streaming connections.

In order to get message reliability, we implement Apache Kafka as the distributor. Messages are generated in Kafka cluster before proceeding to the streaming engine. Each message consists of 30–35 words that has record size for about 5–10 bytes. It is important to note that degree of parallelism of Storm spout and Spark receiver is dependent on the number of topic partitions from Kafka brokers. In order to get desired degree of parallelism, each topic with different number of partition needs to be set upon implementation.

Our experiments with storm demonstrated significant contribution of the spout and bolt parallelization on a single node. Figure 1 depicts the combination of spout and bolt parallelization resulting in linearly increasing throughput (tuple/sec). It can be seen that spout replication allows more tuple in the stream supporting an increase in traffic and throughput. However, each replication requires more workers as spout replications use most of threads in single node (with 4 executors allocation) because spout component is the first in the topology and thus is first to be served through available slots. This has been

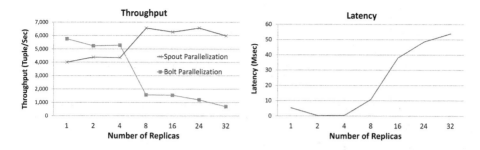

Fig. 1. Measuring Storm throughput and latency on single node.

noticed when degree of parallelism reached 8. Since there are only 4 executors initially, bolt performance decrease dramatically with increased spout replication beyond threads threshold with 8 replicas. Figure 1 also indicates the latency for bolt degree of parallelism processing within single node. It was initially expected that latency will be stable throughout the duration of topology processing. However, we found that, due to overloading threads consumption, latency spikes drastically when bolt gets overloaded. This is because numbers of messages emitted from spout are building up the queue in internal buffer system that bolts access and bolt can only process as much as it is available in the buffer. Note that, it is also due to the fact that spouts have used all the available executors which results in 'overstressed' bolts. This suggests that Storm has a good respond in maintaining the topology under low level of resources. It is evident that there must be a restriction mechanism system similar to Spark to handle such circumstances.

In order to compare one-at-the-time ability for horizontal scaling, we carried out a performance test with Storm on multiple nodes as a comparison with the single node. Figure 2 indicates the horizontal scalability that supports data distribution and load balance through adding more worker nodes. With the increasing of degree of parallelism, more executors are needed to maintain number of threads spawned. Thus, there is need to try to balance the number of

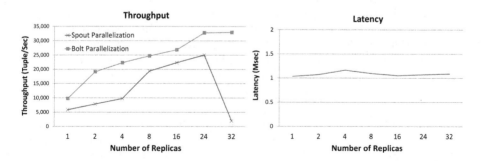

Fig. 2. Measuring Storm throughput and latency based on multiple nodes.

worker nodes with degree of parallelism (e.g. 4 degree of parallelism = 4 worker nodes). As the number of worker nodes increases to match degree of parallelism, throughput is steadily increased because for each increasing input stream in spout it will allow more tuples to be consumed by the topology.

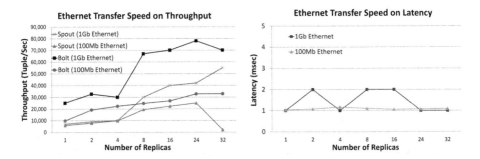

Fig. 3. Measuring Storm Throughput and Latency based on Ethernet switch transfer speed (100 Mb and 1 Gb).

In relation to latency, by adding more workers, it has been proven to reduce the load of internal processing by distributing it across the cluster. This is achieved by sharing incoming tuple to each given workers to reduce bottleneck in the internal buffer. We have identified that by adding more worker nodes decrease overloading of bolts by exploiting more executors, which was also evident from the decrease in bolt capacity. Bolt capacity is measured 0 to 1 - where the capacity over 1 indicates overloading bolt and one should increase the degree of parallelism. Note that in part this is also due to data distribution across nodes which help Storm reducing excessive tuple queues in the internal buffer. Furthermore, it is worth to note that Storm internal processing (bolt) is powerful to handle huge amount of streaming data and then process it with very low latency less than 1 ms.

In terms of the impact of network protocol, since messages are delivered from socket to socket through TCP/UDP connection protocol, we expected that improvement in transport layer will affect the message deliveries. This was indeed the case. Figure 2 demonstrates that with 100Mb Ethernet switch, our cluster has reached the threshold when the *degree of parallelism* was 24, but with faster network (1Gb) threshold was 32. Figure 3 indicates not only improvement in level of threshold, but also in overall performance for both throughput and latency (Fig. 4).

Our experiments have shown a significant *throughput* improvement from micro-batch compared to Storm. It may be noted that our prior experience has suggested insignificant differences in using different micro-batch interval size; the results are sufficiently general with the case of 1 second batch interval. Result shows that *degree of parallelism* does not significantly benefit neither *throughput* nor *latency*. Spark Streaming performance on multiple nodes is worse compared

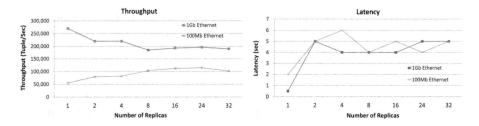

Fig. 4. Measuring Spark throughput and latency based on multiple nodes.

to the single node performance with default degree of parallelism configuration. In comparison to Storm, on single node, Spark is able to perform 15 times better in terms of throughput and on average 4 times faster on multiple nodes. However, the trade-off from having high throughput is the increasing of latency into seconds. Note that latency on the order of seconds is not sufficient for processing real-time analysis. Our experiment also shows that Spark Streaming will randomly choose worker nodes for the scheduler task. A drawback from this approach is that each opening and closing connection between master - workers will increase network traffic which affects the latency.

Figure 5 highlights the comparison between Storm and Spark based on maximum throughput and latency. Note that on Storm we measure the throughput based on spout as it represents how many tuples have been processed inside the topology. We performed several tests on different Spark micro-batching interval from 2 to 0.1 s. It is evident that the micro-batch interval does not affect the maximum throughput (calculated in messages per second) or latency. Spark Streaming does not benefit from degree of parallelism on multiple worker nodes.

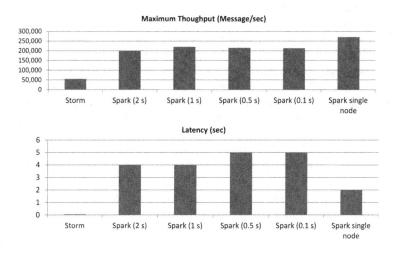

Fig. 5. The comparison between Storm and Spark in Throughput and Latency.

Additionally, Spark has the restriction mechanism that will not allow for total number of executors to be less than the degree of parallelism, for example on a single node with 4 cores one can not have more than 4° of parallelism. On the other hand, in Storm, it would be possible to have 8° of parallelism on the 4 core node, but this will put the workers into an overloading state. Since Spark core framework exploits main memory, its mini-batch processing can appear as fast as one at a time processing adopted in Storm, in spite the fact that the RDD units are larger than Storm tuples.

The benefit from mini-batch is to enhance the throughput in internal engine by reducing data shipping overhead such as lower overhead for ISO/OSI transport layer header which will allow the threads to concentrate on computation [3]. However, one may argue that the consequence of batching concept will add additional latency compared to one at a time. Since the processing of message transfer between worker nodes is performed in network transport layer, message transfer in batch will decrease network I/O. Thus, this will directly reduce CPU consumption and network bandwidth.

4   Conclusion and Future Work

In this work, we evaluated the performance of streaming engines as a variable of their degree of parallelism and number of worker nodes. We showed that by choosing an appropriate degree of parallelism in combination with the right number of worker nodes one can achieve better performance. For example in the case of 20 cluster nodes with four executors on each node optimal is 24° of parallelism, as after 24 the performance will drop.

We have also demonstrated that network bandwidth will increase performance of streaming engines and the level of threshold, however, also after the threshold the performance drops.

Another finding was that Storm has lower *latency* (in milliseconds) and lower *throughput* both on single and multiple nodes compared to Spark. Note that in Storm architecture, *throughput* can be improved by adding more worker nodes to allow more *degree of parallelism*. Spark micro-batch processing delivers high *throughput* with high *latency* penalty. In our experiments, Spark had better *throughput* compared to Storm (15 times on single worker node and 4 times on multiple worker nodes with 4 executors on each node) but Storm delivered better latency over Spark.

In our future work, we plan to investigate the impact of configuring a different number of *tasks*, which in these experiments was not explicitly defined, rather it was left to engine to select them. Second, we would like to explore the impact of *data locality* on performance. We would also like to consider the impact of the maximum tuples that a spout will allow to be processed in the topology on engine performance.

References

1. Bedini, I., Sakr, S., Theeten, B., Sala, A., Cogan, P.: Modeling performance of a parallel streaming engine: Bridging theory and costs. In: Proceedings of the 4th ACM/SPEC International Conference on Performance Engineering, pp. 173–184. ICPE 2013, NY, USA. ACM, New York (2013). http://doi.acm.org/10.1145/2479871.2479895
2. Casale, G., Ustinova, T.: State of the art analysis (2015)
3. Lohrmann, B., Janacik, P., Kao, O.: Elastic stream processing with latency guarantees (2015)
4. Lohrmann, B., Warneke, D., Kao, O.: Nephele streaming: stream processing under qos constraints at scale. Cluster Comput. **17**(1), 61–78 (2014). http://dx.doi.org/10.1007/s10586-013-0281-8
5. Neumeyer, L., Robbins, B., Nair, A., Kesari, A.: S4: Distributed stream computing platform. In: 2010 IEEE International Conference on Data Mining Workshops (ICDMW), pp. 170–177, December 2010
6. da Silva Morais, T.: Survey on frameworks for distributed computing: Hadoop, spark and storm (2015)
7. Theeten, B., Bedini, I., Cogan, P., Sala, A., Cucinotta, T.: Towards the optimization of a parallel streaming engine for telco applications. Bell Labs Techn. J. **18**(4), 181–197 (2014)
8. Toshniwal, A., Taneja, S., Shukla, A., Ramasamy, K., Patel, J.M., Kulkarni, S., Jackson, J., Gade, K., Fu, M., Donham, J., et al.: Storm@ twitter. In: Proceedings of the 2014 ACM SIGMOD international conference on Management of data, pp. 147–156. ACM (2014)
9. Zaharia, M., Chowdhury, M., Das, T., Dave, A., Ma, J., McCauley, M., Franklin, M.J., Shenker, S., Stoica, I.: Resilient distributed datasets: A fault-tolerant abstraction for in-memory cluster computing. In: Proceedings of the 9th USENIX conference on Networked Systems Design and Implementation, pp. 2–2. USENIX Association (2012)
10. Zaharia, M., Das, T., Li, H., Hunter, T., Shenker, S., Stoica, I.: Discretized streams: Fault-tolerant streaming computation at scale. In: Proceedings of the Twenty-Fourth ACM Symposium on Operating Systems Principles, pp. 423–438. ACM (2013)

BigDap 2016 – Big Data Applications and Principles

Reducing Big Data by Means
of Context-Aware Tailoring

Paolo Garza[1], Elisa Quintarelli[2], Emanuele Rabosio[2(✉)], and Letizia Tanca[2]

[1] Politecnico di Torino, Corso Duca Degli Abruzzi 24, 10129 Torino, Italy
paolo.garza@polito.it
[2] Politecnico di Milano, Piazza Leonardo Da Vinci 32, 20133 Milano, Italy
{elisa.quintarelli,emanuele.rabosio,letizia.tanca}@polimi.it

Abstract. *Context-aware personalization* is one of the possible ways
to face the problem of *information overload*, that is, the difficulty of
understanding an issue and making decisions when receiving too much
information. Context-aware personalization can *reduce the information
noise*, by proposing to the users only the information which is relevant
to their current contexts. In this work we propose an approach that
uses data mining algorithms to automatically infer the subset of data
that, for each context, must be presented to the user, thus reducing the
information noise.

Keywords: Contextual views · Association rules · Data tailoring

1 Introduction

In this Big Data era, we are surrounded by huge collections of data, be it from
information systems, documents, or the Internet; besides, the support of the
currently available mobile devices grants this data *high portability*, thus allowing
users to access the information they need at any time and place. Anyhow, for a
user, extracting just the useful information from all the data that is irrelevant
to her specific situation may be long and difficult, and thus all this richness may
generate confusion instead of value.

It is by now commonly agreed [1,2,6,11] that a way to effectively manage
the big data phenomenon is *contextualization*, that is, providing people, at any
place and time, only with the portions of data that can be useful in that particular moment. However, automatizing the process of distinguishing, moment
by moment, the useful data from all the information which is irrelevant to the
specific application context or user is not a trivial task. For example, if we consider a movie dataset and a family with young children interested in movies, a
contextual system will propose the movies played in cinemas close to the family
location and appropriate for children.

This research is partially supported by the IT2Rail project funded by European
Union's Horizon 2020 research and innovation program under grant agreement No:
636078, and by the grant "IBM – International Business Machine – 2014".

© Springer International Publishing Switzerland 2016
M. Ivanović et al. (Eds.): ADBIS 2016, CCIS 637, pp. 115–127, 2016.
DOI: 10.1007/978-3-319-44066-8_13

The literature has coped with this research challenge by proposing context models (see [1, 2, 6] for surveys). Such models allow the reduction of large datasets on the basis of some perspectives (dimensions) describing the situation (i.e., the *context*) in which the user is involved; typical such dimensions are, for example, the user's role and her location. More sophisticated context parameters can also be introduced, like the current activity of the user, or her main interest topic. We call the activity of selecting the relevant information for a target application or user, in a specific context, *data tailoring* [4]: *data tailoring* refers to the capability of the system to provide the users only with the view (over a global, centralized or not, database) that is relevant for their current context. However, with a large variety of data and a considerable number of possible contexts, the manual specification of contextual views may be a trying experience, discouraging the designer; moreover, such an approach cannot take into account the evolution of user tastes or situations, which may generate, during the system life, changes in the user interests and needs[1]. As a way around this problem, considering relational databases, we propose to apply data mining algorithms to the past querying activity of the user to learn which data are appropriate for each possible context, and then study two alternative techniques. To the best of our knowledge, this is the first proposal addressing the automatic extraction of context-aware views.

The paper is structured as follows. Section 2 describes the movie scenario used in the examples and briefly introduces the context model and the definition of contextual views used throughout the paper, while Sect. 3 contains our proposals to infer contextual views by means of data mining algorithms. Section 4 discusses preliminary experiments and, finally, Sect. 5 draws the conclusions.

2   Background

The Running Example. Throughout the paper we use as running example in the movie domain, storing information about movies and actors, composed of three tables: MOVIE, ACTOR, MOVIE_ACTOR. Figure 1 contains a sample instance of the database.

MOVIE			
movie_id	title	genre	year
m1	Movie1	comedy	2008
m2	Movie2	thriller	2011
m3	Movie3	thriller	2008

ACTOR		
actor_id	name	citizenship
a1	Actor1	Italian
a2	Actor2	Italian
a3	Actor3	French

MOVIE_ACTOR	
movie_id	actor_id
m1	a1
m1	a3
m2	a2
m2	a3
m3	a2

Fig. 1. The three tables of our running example

The Context Model. The Context Dimension Model [4] provides the constructs to design the *context schema* (or *Context Dimension Tree – CDT*) of a

---

[1] *Context evolution* [10] is the research topic that takes this into account; however, if this task is performed by the designer, it makes his or her burden even heavier.

given scenario (see Fig. 2 for the CDT of our running example) by means of a hierarchical structure consisting of (i) *context dimensions* (black nodes), modeling the perspectives through which the user perceives the application domain (e.g. *user, int_topic, situation, time,* and *day*) and (ii) the envisaged *dimension values* (white nodes), i.e., the values used to select the context-aware information (e.g. the users `adult`, `teenager` and `familyWithChildren`).

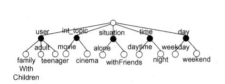
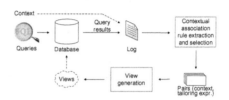

Fig. 2. The Context Dimension Tree for the running example

Fig. 3. The extraction of the contextual views

A set of dimension values from the CDT represents a possible user context, and, from a formal point of view, a context is defined as a conjunction of *context elements* (statements of the form *<dimension=value>*). For instance, `situation` = `withFriends` ∧ `time` = `daytime` is one of the contexts associated with the CDT reported in Fig. 2.

Design-time Definition of Contextual Views. The work [4] proposes the CARVE methodology to address context-aware view definition. The CARVE methodology requires that the designer specifies the context schema at design-time, identifying the possible contexts the user may find herself in at run-time. Moreover, also the tailoring of relevant information is manually specified at design time. Indeed, the designer must associate each context with a view on the relational database. This association remains *virtual*, and the views will be materialized at run time, when the system presents the user only with data useful for her currently active context. Referring to the relational data model, a contextual view corresponds to one or more relational views over a global schema, in general to a set of views; in the following, by the general word view we will thus indicate a set of (virtual) relations.

Given the relational schema $\mathcal{D}$ of the target scenario, the context-aware view for context C is expressed as a set of *tailoring expressions*, one for each $R \in \mathcal{D}$.

A *tailoring expression* e^R on a relation R can be used to: (i) pick the entire relation R, (ii) combine different expressions on R by means of the union operator, (iii) filter some tuples by specifying a selection condition and/or by using the semi-join operator with other relations. Note that tailoring expressions do not include the projection operator, and thus the schema of the view associated with each relation R is equal to that of R in the original database. With this restriction the tailoring process returns a database that is reduced only in terms of tuples. Given a tailoring expression e^R over R, R is called *origin* of e^R.

A *contextual view* is then a set of tailoring expressions, one for each relation of the relational schema R.

3   Mining Contextual Views

Differently from previous work [3], the novelty of the research we present here is that the generation of the contextual view of each context and the association between a context and the corresponding contextual view are performed in a fully automatic way, starting from datasets obtained by means of data mining. The automatic extraction of contextual view definitions is a novel issue, that has never been investigated before. In this section we propose two alternative solutions for this problem, both relying on association rule mining.

With the same final objective, i.e., the extraction of the definitions of the contextual views, the two strategies we propose proceed through the same phases, summarized in Fig. 3. They require as input the access log for each table of the database, containing the results of the queries which have involved that table. More precisely, each relation schema $R \in \mathcal{D}$ is associated with a log LOG_R, that is a list containing entries of the form *(context, primary key)*. Every time the user issues a query in a certain context C, for each tuple of R with key pk involved in the query answer, a pair (C, pk) is added to LOG_R. The log is then used to mine *contextual association rules*, putting into relation contexts with features of the data; then, the most relevant ones are selected in order to generate a model M_C for each context C of interest. The mining and selection steps are performed in different ways by the two strategies, each of which produces as output a set $\mathcal{S}$ of pairs *(context, tailoring expression)*; note that each context may well be associated with several tailoring expressions. Finally, the set $\mathcal{S}$ is used to generate the contextual views.

The two methodologies approach the discovery of contextual views following two opposite philosophies. The first one, the *Context-To-View* algorithm, searches rules of the form *context* → dataset, with the aim of learning which data features have been the most popular in the various contexts. The second approach, *View-To-Context*, tries to discover rules in the form *dataset* → context, and is a little more fine-grained: this strategy starts from the data, trying to understand whether certain data are or are not relevant to the different contexts.

In the rest of this section we describe in detail the two alternative algorithms (Subsects. 3.1 and 3.2), and then explain how their outputs are used to derive the actual contextual views (Subsect. 3.3).

3.1   Context-To-View Algorithm

The first solution that we present, named Context-To-View, is inspired by the PREMINE methodology [9], whose aim is the extraction of quantitative contextual preferences from the logs of the past data accesses of each individual user. The automatic mining of quantitative contextual preferences is based on the use of association rules of the form $C \rightarrow e^R$, where C is a valid context and e^R is a tailoring expression over R, not containing unions.

The extracted rules represent relationships between contexts and data; based on the extracted rules, a model, dubbed M_C, is generated for each context C of interest by selecting the association rules having C as antecedent. As described in the following, the generated models are used to populate the output set $\mathcal{S}$, i.e., the set of pairs (*context, tailoring expression*) used to define contextual views. The subset of association rules to include in each model M_C is based on a filtering step that exploits a quantitative score. To generate the contextual models for relation R, the Context-To-View algorithm operates as follows:

1. Analyze the log table LOG_R related to R and generate a transaction log $TLOG_R$. The transaction log is needed in order to extract (contextual) association rules.
2. Mine frequent rules of the form $C \rightarrow e^R$ by means of an FP-growth-based association rule mining algorithm.
3. Select high quality rules to be included in the contextual models by applying a filter based on the score (see below) of the mined rules. A model M_C is generated for each context C by considering the subset of rules with C as antecedent.
4. Use the consequents of the selected rules to define the output set $\mathcal{S}$ exploited to determine the contextual views.

In the following, each step of Context-To-View is described in detail.

Contextual Association Rule Extraction. The first step of Context-To-View mines the contextual association rules of the form $C \rightarrow e^R$. In order to extract the rules, the log table LOG_R related to R must be converted into a set of transactions, as usual in the scope of association rules. A transaction is added for each tuple of LOG_R, and contains items related to the context, the values of the attributes of R and those of the attributes of the relation schemas reachable from R using foreign keys.

Itemset mining is then performed with FP-growth [5], appropriately modified to prevent the extraction of itemsets that cannot lead to rules in the form $C \rightarrow e^R$. Then, association rules with the required features are generated from the itemsets.

Example 1. Suppose that some users have accessed the database in Fig. 1, generating the following log for the MOVIE table:

- $(\texttt{situation} = \texttt{withFriends} \wedge \texttt{time} = \texttt{daytime}, m1)$
- $(\texttt{situation} = \texttt{withFriends} \wedge \texttt{time} = \texttt{daytime}, m3)$
- $(\texttt{situation} = \texttt{withFriends} \wedge \texttt{time} = \texttt{daytime}, m1)$
- $(\texttt{situation} = \texttt{withFriends} \wedge \texttt{time} = \texttt{night}, m2)$

Figure 4 shows $TLOG_{\text{MOVIE}}$ in transaction format, ready for the mining process, and Fig. 5 lists the association rules in form $C \rightarrow e^R$ that can be extracted using $minsup = 0.1$ and $minconf = 0.4$. For brevity, only the rules with just one item in the consequent and with the contexts situation = withFriends $\wedge$ time = daytime and situation = withFriends in the antecedent are considered; moreover, the movie title and the actor name are excluded from the mining, since they are usually in a one-to-one relationship with the movie and actor identifiers.

– situation=withFriends, time=daytime, movie_id='m1', genre='comedy', year='2008', actor.actor_id='a1', actor.citizenship='Italian', actor.actor_id='a3', actor.citizenship='French'
– situation=withFriends, time=daytime, movie_id='m3', genre='thriller', year='2008', actor.actor_id='a2', actor.citizenship='Italian'
– situation=withFriends, time=daytime, movie_id='m1', genre='comedy', year='2008', actor.actor_id='a1', actor.citizenship='Italian', actor.actor_id='a3', actor.citizenship='French'
– situation=withFriends, time=night, movie_id='m2', genre='thriller', year='2011', actor.actor_id='a2', actor.citizenship='Italian', actor.actor_id='a3', actor.citizenship='French'

Fig. 4. Transaction log associated with MOVIE

	Rule	Confidence
R1	situation=withFriends $\wedge$ time=daytime $\longrightarrow$ genre='comedy'	2/3
R2	situation=withFriends $\wedge$ time=daytime $\longrightarrow$ year='2008'	1
R3	situation=withFriends $\wedge$ time=daytime $\longrightarrow$ movie_id='m1'	2/3
R4	situation=withFriends $\wedge$ time=daytime $\longrightarrow$ actor.actor_id='a1'	2/3
R5	situation=withFriends $\wedge$ time=daytime $\longrightarrow$ actor.actor_id='a3'	2/3
R6	situation=withFriends $\wedge$ time=daytime $\longrightarrow$ actor.citizenship='Italian'	1
R7	situation=withFriends $\wedge$ time=daytime $\longrightarrow$ actor.citizenship='French'	2/3
R8	situation=withFriends $\longrightarrow$ genre='comedy'	1/2
R9	situation=withFriends $\longrightarrow$ genre='thriller'	1/2
R10	situation=withFriends $\longrightarrow$ year='2008'	3/4
R11	situation=withFriends $\longrightarrow$ actor.actor_id='a1'	1/2
R12	situation=withFriends $\longrightarrow$ actor.actor_id='a2'	1/2
R13	situation=withFriends $\longrightarrow$ actor.actor_id='a3'	3/4
R14	situation=withFriends $\longrightarrow$ actor.citizenship='Italian'	1
R15	situation=withFriends $\longrightarrow$ actor.citizenship='French'	3/4

Fig. 5. Some rules extracted from the log of Example 1 by Context-To-View

Selecting Relevant Rules. The mined rules associate contexts with tailoring expressions, and are characterized by a confidence that should provide a quantitative measure of the strength of the association. However, as discussed in [7,9], a high confidence for a rule does not necessarily indicate a strong preference of the users. Indeed, the extent of the preference depends also on the accessibility of the data mentioned in the rule: even if some features of the data, such as a particular movie genre, are not liked by the users, they can be recurring in the log if they occur very frequently in the database and, conversely, interesting but uncommon features may obtain low confidence.

Consequently, the PREMINE methodology evaluates the strength of a rule relying not only on its confidence in the log, but also on the frequency of the related data in the database. In this work we adopt the same formula as defined in PREMINE [9], computing the following score in the range (0.5, 1] for an association rule r:

$$score(r) = \min((1 + \gamma \cdot \frac{confidence(r)}{frequency(r.consequent)}) \cdot 0.5, 1) \quad \text{if} \quad \frac{confidence(r)}{frequency(r.consequent)} > 1 \quad (1)$$

The parameter γ is used to tune the contribution of the ratio between the confidence of the rule under analysis and the frequency of its consequent to the score, with the aim of avoiding the production of too many scores equal to 1. The most suitable value for γ depends on the specific application, and its choice is left to the designer. According to Eq. (1), the higher the difference between the rule confidence and the frequency of its consequent, the "stronger" the rule. If the confidence of r is less than or equal to the frequency of its consequent, the rule is discarded. Further details on the rationale underlying Eq. (1) can be found in [9].

The computation ends with a set of association rules, each putting in relation a context with a tailoring expression, associated with the score resulting from Eq. (1). In order to select the rules considered relevant, a suitable minimum score threshold min_score is applied on this score. The selected rules are used to generate contextual models. More specifically, a model M_C is generated for each context C. Based on the generated models, an arbitrary tuple t is considered of interest for context C if t satisfies all the conditions in the consequent of at least one of the rules in M_C. Note that each tuple t can be associated with many contexts because t can satisfy the rules of several models M_C. The output set S is filled by generating one pair *(context, tailoring expression)* for each rule.

For instance, consider the rules R1–R7 in Fig. 5, referred to the context situation = withFriends $\wedge$ time = daytime. The frequency of the conditions $genre =$ '*comedy*', $movieid =$ '*m1*' and $actor.actorid =$ '*a1*' in the database is equal to $\frac{1}{3}$, while that of $year =$ '*2008*', $actor.actorid =$ '*a3*' and $actor.citizenship =$ '*French*' is $\frac{2}{3}$; the frequency of $actor.citizenship =$ '*Italian*', instead, is equal to 1. As a consequence, rules R5, R6 and R7 are excluded, because their confidence is not greater than the associated frequency. For the remaining rules, Eq. (1) is employed to compute the score. Choosing $\gamma = 0.3$, the score obtained for R1, R3 and R4 is 0.8, while that computed for R2 is 0.725. Applying a minimum score threshold $min\_score = 0.75$ to select the relevant rules, R2 is excluded. It follows that the model for context situation = withFriends $\wedge$ time = daytime is composed of rules R1, R3, and R4 and hence the set S that is output, considering only the table MOVIE in the context situation = withFriends $\wedge$ time = daytime denoted by C, and writing the tailoring expressions in relational algebra, is as follows: $S = \{(C, \sigma_{genre=\text{'comedy'}}\text{MOVIE}), (C, \sigma_{movie\_id=\text{'m1'}}\text{MOVIE}), (C, \text{MOVIE} \bowtie \sigma_{actor\_id=\text{'a1'}}\text{MOVIE\_ACTOR})\}$.

Note that other pairs will be included in S for the other contexts by considering the associated models.

3.2 View-To-Context Algorithm

Given an arbitrary context C, View-To-Context infers from the query log a model that is used to predict if a new tuple t is relevant to C or not, based on the values of its attributes. The model is based on a set of association rules of the form $e^R \rightarrow C$, where C is a valid context and e^R is a tailoring expression without unions over relation R. The mined rules allow representing relationships between data and contexts.

The basic intuition that is exploited by View-To-Context to infer the model M_C consists in automatically extracting the characteristics (i.e., conjunctions of conditions $attribute_i = value_i$) that are frequent in the tuples already accessed in context C but infrequent in the tuples that have not been accessed in context C. These characteristics can be used to model the data relevant to C.

Given a relation R, View-To-Context builds a set of models M_C. Specifically, during the model generation phase, View-To-Context considers one context at a time and applies the following steps in order to generate M_C:

1. Analyze the log of the tuples of relation R accessed in context C and create a transaction log specific for context C, $TLOG_R^C$.
2. Mine frequent contextual rules of the form $e^R \to C$ by means of an FP-growth-based association rule mining algorithm.
3. Select the rules to be included in the generated model by applying a selection based on a database coverage technique.

   These steps are described in detail below.

Contextual Association Rule Extraction. The first two steps of View-To-Context (i) generate a transactional log containing the data to be analyzed and (ii) mine (contextual) association rules from the transaction log, respectively. The transaction logs generated by View-To-Context are derived from the various LOG_R, i.e., the log of the accessed tuples of relation R. Specifically, View-To-Context generates one transaction log for each context and performs a rule mining process for each context.

The transaction log $TLOG_R^C$ associated with relation R and context C is generated as follows:

– For each entry in LOG_R, associated with context C and a tuple $r \in R$, one transaction is inserted in $TLOG_R^C$. The transaction is composed of all pairs *(attribute_i, value_i)* generated from tuple r and the pairs *(attribute_j, value_j)* associated with the tuples of other relations R' reachable from $r \in R$ through foreign keys. The generated transaction is associated with context C. This step is similar to the construction of $TLOG_R$ in Context-To-View.
– For each entry in LOG_R, associated with a context different from C and a tuple $r \in R$ that has *never* been accessed at least once in context C, one transaction is inserted in $TLOG_R^C$. Also in this case, the transaction is composed of all pairs *(attribute_i, value_i)* generated from tuple r and the pairs *(attribute_j, value_j)* associated with the tuples of other relations R' reachable from $r \in R$ through foreign keys. The generated transaction is without context because it is not associated with the context C the algorithm is currently analyzing.

(situation=withFriends ∧ time=daytime)	movie_id='m1', genre='comedy', year='2008', actor.actor_id='a1', actor.citizenship='Italian', actor.actor_id='a3', actor.citizenship='French'
(situation=withFriends ∧ time=daytime)	movie_id='m3', genre='thriller', year='2008', actor.actor_id='a2', actor.citizenship='Italian'
(situation=withFriends ∧ time=daytime)	movie_id='m1', genre='comedy', year='2008', actor.actor_id='a1', actor.citizenship='Italian', actor.actor_id='a3', actor.citizenship='French'
	movie_id='m2', genre='thriller', year=' 2011', actor.actor_id='a2', actor.citizenship='Italian', actor.actor_id='a3', actor.citizenship='French'

Fig. 6. Transaction log associated with relation MOVIE and context `situation = withFriends ∧ time = daytime`

	Rule	Confidence	Support
R1	year='2008' → situation=withFriends ∧ time=daytime	1	3/4
R2	genre='comedy' → situation=withFriends ∧ time=daytime	1	1/2
R3	movie_id='m1' → situation=withFriends ∧ time=daytime	1	1/2
R4	actor.actor_id='a1' → situation=withFriends ∧ time=daytime	1	1/2
R5	actor.actor_id='a1' ∧ year=' 2008' → situation=withFriends ∧ time=daytime	1	1/2
R6	actor.citizenship='Italian' → situation=withFriends ∧ time=daytime	3/4	1
R7	actor.actor_id='a3' → situation=withFriends ∧ time=daytime	2/3	3/4
R8	actor.citizenship='French' → situation=withFriends ∧ time=daytime	2/3	3/4
...	...	...	...

Fig. 7. Some rules extracted from the log of Example 2 by View-To-Context

Example 2. Figure 6 shows the transaction log generated for context `situation = withFriends ∧ time = daytime` from the LOG_{MOVIE} of Example 1. As you can notice the transactions related to movies 'm1' and 'm3' are associated with context `situation = withFriends ∧ time = daytime` because both movies have been accessed in that context. Specifically, 'm1' appears twice because it is present in two rows of the log of table MOVIE, while 'm3' appears only once. Differently, 'm2' is not associated with context `situation = withFriends ∧ time = daytime` because it has never been accessed in the considered context.

From the transaction log, rules of the form $e^R \rightarrow C$ are mined by means of an FP-growth-based mining algorithm. However, only the rules having context C as consequent are considered in the following steps to generate the model M_C. For example, Fig. 7 reports some of the rules that have been mined from the transaction log reported in Fig. 6 by setting $minsup = 0.33$ and $minconf = 0$.

Selecting Relevant Rules. The output of the mining step is a set of rules representing the relationship between data, in terms of conjunctions of pairs (attribute, value), and context. However, some of them are redundant. For this reason a database coverage approach, similar to that normally used by the associative classifiers [8], is used. The database coverage approach selects a minimal set of high quality rules able to "cover/represent" the tuples accessed in the context under analysis. More specifically, it greedily selects the minimal number of rules needed to have in the final model M_C at least one matching rule for every tuple accessed (at least one time) in context C. An arbitrary rule r matches an arbitrary tuple t iff all the conditions in the antecedent of r are satisfied by t.

The greedy database coverage step operates as follows:

– Consider a rule r at a time in decreasing confidence and decreasing support order.

- Select the subset of tuples $matched_C$, defined as the tuples that have been accessed at least once in context C and are matched by r.
- Select the subset of tuples $matched_{\neg C}$, defined as the tuples that have never been accessed in context C and are matched by r.
- If the cardinality of $matched_C$ is equal to or greater than 1 and the cardinality of $matched_{\neg C}$ is lower than a maximum error threshold max_err then (i) r is included in model M_C and (ii) the tuples in $matched_C \cup matched_{\neg C}$ are excluded in the following iterations.

The process ends when all rules have been considered.

The max_err threshold is the only parameter of the database coverage step. It is used to avoid the selection of low quality rules. More specifically, a rule is a low quality rule if it matches too many tuples that have never been accessed in context C (i.e., the conditions in its antecedent are not able to represent the data frequently accessed in context C).

For instance, consider the following rule:

$actor.actor\_id = 'a1' \rightarrow situation = withFriends \wedge time = daytime$, confidence $= 100\,\%$.

According to this example rule, if actor 'a1' is part of the cast of movie m then, with a probability $100\,\%$, m is of interest when the context is situation $=$ withFriends $\wedge$ time $=$ daytime.

Differently from the Context-To-View algorithm, the View-To-Context algorithm mines rules with the conditions used to identify the tailoring expressions in the antecedent of the rules and a context in the consequent. Since the confidence measure tends to be biased towards rules with frequent events in their consequent, the Context-To-View algorithm used a more sophisticated measure to score rules, combing confidence and frequency of their consequent, to avoid the mentioned bias given by frequent events. View-To-Context exploits the confidence, and no other measure, because in this case we want to rank rules with the same consequent and hence the confidence measure is able to accomplish our goal. In fact, a rule has a high confidence only if the tailoring expression in its antecedent matches tuples accessed frequently in context C and infrequently in the other contexts, i.e., the tailoring expression under consideration characterizes data that are interesting exclusively for context C.

Consider now the set of rules reported in Fig. 7, mined from the transaction log in Fig. 6. By applying the database coverage approach on that set of rules, only one rule is included in the model $M_{\text{situation=withFriends}\wedge\text{time=daytime}}$. All the other rules are redundant or not representative of the main characteristics of the context C under analysis. More specifically, only rule R1 is selected. The rule states that all movies made in year 2008 are of interest when the context is situation $=$ withFriends $\wedge$ time $=$ daytime. If you consider the transaction log in Fig. 6, you can notice that this property allows discriminating among movies accessed in context situation $=$ withFriends $\wedge$ time $=$ daytime and movies not accessed in that context.

The association rules composing model M_C are used to generate the pairs (C, *tailoring expression*) to include in the output set S for context C. More specifically one pair is included for each rule in M_C, by using the antecedent of the rule as tailoring expression. For instance, the output set $\mathcal{S}$ of View-To-Context, considering only context $\mathtt{situation} = \mathtt{withFriends} \wedge \mathtt{time} = \mathtt{daytime}$, denoted by C, and the MOVIE table is as follows: $\mathcal{S} = \{(C, \sigma_{\text{year}='2008'}\text{MOVIE})\}$.

Note that usually a subset of rules is selected. In this case only one rule has been selected because the log considered in the running examples is very small.

3.3   The Contextual Views

Once the output set $\mathcal{S}$ has been computed (by either algorithm), determining the view to be associated with a certain context is straightforward. Since all the elements of $\mathcal{S}$ represent information that has been judged relevant, the tailoring expression e_C^R describing the contextualized version of a relation schema R in a context C is computed as *the union of all the tailoring expressions in $\mathcal{S}$ that are related to C and whose origin is R.*

Example 3. The final tailoring expressions that in our running example are associated with MOVIE in the context $\mathtt{situation} = \mathtt{withFriends} \wedge \mathtt{time} = \mathtt{daytime}$ with Context-To-View and View-To-Context, respectively, are:

- $\sigma_{\text{genre}='comedy'}\text{MOVIE} \cup \sigma_{\text{movie\_id}='m1'}\text{MOVIE} \cup \text{MOVIE} \bowtie \sigma_{\text{actor\_id}='a1'}\text{MOVIE\_ACTOR}$
- $\sigma_{\text{year}='2008'}\text{MOVIE}$

As this simple example shows, the two algorithms may generate different contextual views.

4   Preliminary Evaluation

A set of preliminary experiments were executed on a dataset containing TV-viewing information related to 4867 movies, broadcast both over the air and by satellite. The dataset is composed of the description of the movies and a log of the movies watched by the users. The log contains 888670 rows, each specifying the identifier of the user and that of the movie she watched, along with the time stamp and the people who are watching the program with that user. The latter two pieces of information were used to determine the values of three context dimensions: day of the week, time slot and situation. We identified five possible relevant values for the situation: adults and children, group of adults, group of children, adult alone, child alone.

To evaluate the quality of the proposed algorithms, we compare the frequency of correct decisions in including tuples in the context-aware views associated with the MOVIE relation, with respect to the size of the automatically defined contextual views. To this aim, the *number of hits* and *view size* measures are used. The number of hits (#hits) for the context C measures how often the movies (tuples in MOVIE) which are needed to answer to the queries executed in context C were actually included in the contextual view associated with C, whereas view size measures the number of tuples included in the view of context

C. To obtain an average value among the possible contexts, we consider the *micro-average(#hits)* and *micro-average(view size)* measures [12].

Figure 4 displays the number of hits with respect to the view size. The first observation that can be drawn is that both the experimented techniques show good values for the number of hits. In addition, the graph highlights how View-To-Context achieves a better accuracy than Context-To-View.

Note that, however, View-To-Context requires to perform the mining once for each context, whereas Context-To-View requires just a single execution of the mining. This makes Context-To-View more efficient, especially in the scenarios characterized by a great number of contexts. Therefore, the designer may choose the algorithm to be used depending on the application requirements: if the context-aware views are stable and the execution time is not important, the more accurate View-To-Context is advised. On the contrary, if the views are very dynamic and require frequent refresh then Context-To-View might be the best option.

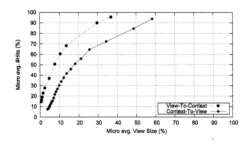

Fig. 8. Number of hits as a function of view size

5   Conclusions

This work has proposed two techniques to automatically infer contextual views from log data. To the best of our knowledge, this is the first attempt to tackle such an issue. Our study, however, is still preliminary. In the future, we plan to deepen our research developing a full-fledged algorithm and to carry out a complete experimental evaluation.

References

1. Baldauf, M., Dustdar, S., Rosenberg, F.: A survey on context-aware systems. Int. J. Ad Hoc Ubiquit. Comput. **2**(4), 263–277 (2007)
2. Bolchini, C., Curino, C., Quintarelli, E., Schreiber, F.A., Tanca, L.: A data-oriented survey of context models. SIGMOD Rec. **36**(4), 19–26 (2007)
3. Bolchini, C., Curino, C., Quintarelli, E., Schreiber, F.A., Tanca, L.: Context information for knowledge reshaping. Int. J. Web Eng. Technol. **5**(1), 88–103 (2009)

4. Bolchini, C., Quintarelli, E., Tanca, L.: CARVE: context-aware automatic view definition over relational databases. Inf. Syst. **38**(1), 45–67 (2013)
5. Han, J., Pei, J., Yin, Y., Mao, R.: Mining frequent patterns without candidate generation: a frequent-pattern tree approach. Data Min. Knowl. Discov. **8**(1), 53–87 (2004)
6. Hong, J., Suh, E., Kim, S.: Context-aware systems: a literature review and classification. Expert Syst. Appl. **36**(4), 8509–8522 (2009)
7. Jung, S.Y., Hong, J.-H., Kim, T.-S.: A statistical model for user preference. IEEE Trans. Knowl. Data Eng. **17**(6), 834–843 (2005)
8. Liu, B., Hsu, W., Ma, Y.: Integrating classification and association rule mining. In: Proceedings of KDD, pp. 80–86. AAAI Press (1998)
9. Miele, A., Quintarelli, E., Rabosio, E., Tanca, L.: A data-mining approach to preference-based data ranking founded on contextual information. Inf. Syst. **38**(4), 524–544 (2013)
10. Quintarelli, E., Rabosio, E., Tanca, L.: A principled approach to context schema evolution in a data management perspective. Inf. Syst. **49**, 65–101 (2015)
11. Stefanidis, K., Pitoura, E., Vassiliadis, P.: Managing contextual preferences. Inf. Syst. **36**(8), 1158–1180 (2011)
12. Sun, A., Lim, E.: Hierarchical text classification and evaluation. In: Proceedings of ICDM, pp. 521–528. IEEE (2001)

Feature Ranking and Selection for Big Data Sets

Bruno Ordozgoiti$^{(\boxtimes)}$, Sandra Gómez Canaval, and Alberto Mozo

Department of Computer Systems, University College of Computer Science,
Universidad Politécnica de Madrid, Crta. de Valencia km. 7, 28031 Madrid, Spain
{bruno.ordozgoiti,sgomez}@etsisi.upm.es, a.mozo@upm.es

Abstract. The availability of big data sets has led to the successful application of machine learning and data mining to problems that
were previously unsolved. The use of these techniques, though, is rarely
straightforward. High dimensionality is often one of the main obstacles
that must be overcome before learning an adequate model or drawing
useful conclusions from large amounts of data. Rank revealing matrix
factorizations can help in addressing this problem, by permuting the
columns of the input data so that linearly dependent and thus redundant ones are moved to the right. These factorizations, however, are
designed to operate in a centralized fashion, requiring the input data to
be loaded into main memory, which makes them inapplicable to large
data sets. In this paper we prove that data sets comprised of a huge
number of rows can be easily transformed into a compact square matrix
that preserves the permutation yielded by rank revealing QR factorizations. This leads to a simple algorithm for running these factorizations
on big data sets regardless of their number of rows. The nature of the
transformation makes it also possible to deal with high dimensional data
with a controlled loss of precision. We offer experimental results showing
that our method can provide improvements for the k-means algorithm,
both in clustering results and in running time.

Keywords: Feature selection · Unsupervised learning · Big data

1 Introduction

Over the last few years, the increasing availability of large amounts of data has
allowed for unprecedented success stories in machine learning and data mining.
Data sets, however, are not always easy to exploit for various reasons. First,
data are often high dimensional, making it difficult to learn machine learning
models and make useful inferences. Second, in certain domains data sets reach
formidable sizes and it is costly to process them. Finally, big data sets are usually
impossible to label, ruling out the application of supervised learning methods.

There exist different methods to overcome the issue of high dimensionality.
Among these, matrix factorization techniques provide a successful, theoretically
well-founded framework for reducing the dimensions of a matrix while preserving
most of its information. Rank-revealing QR (RRQR) factorizations, also known

© Springer International Publishing Switzerland 2016
M. Ivanović et al. (Eds.): ADBIS 2016, CCIS 637, pp. 128–136, 2016.
DOI: 10.1007/978-3-319-44066-8_14

as QR with column pivoting, are an example of these techniques. In addition to the orthogonal and upper-trapezoidal matrices Q and R, they find a permutation such that most of the action in the matrix is retained by the first columns. As such, these factorizations translate naturally to feature selection in unsupervised settings, i.e. for unlabelled data sets.

There exist well-known, highly optimized implementations of RRQR factorizations [1]. These, however, are designed to process matrices that are loaded in main memory, which makes them inapplicable to the huge data sets that arise in many domains today.

In this paper we propose an algorithm to help overcome the issue mentioned above. In particular, we offer a method that enables the execution of RRQR factorizations on data sets comprised of an arbitrary number of rows. For a matrix $A \in \mathbb{R}^{m \times n}$ with $m \gg n$, we prove that a compact representation can be obtained such that the traditional algorithm for the QR factorization with column pivoting yields the same permutation as if run on the whole data set. This results in an unsupervised feature selection algorithm that can be easily implemented relying on well-established, readily available tools, and run on data sets of an arbitrary number of rows. The form of the resulting matrix also makes it possible to deal with large values of n with small or no loss in precision. We conduct experiments and show that this algorithm can improve both the running time and the accuracy of a well-known clustering algorithm.

The rest of the paper is structured as follows: Sect. 2 describes related work on feature selection algorithms. Section 3 describes the contribution of this paper. Section 4 provides a description of the algorithm and the mentioned proof. In Section 5 we present experimental results and we conclude in Sect. 6.

2   Related Work

Rank-revealing QR (RRQR) factorizations emerged as a means to detect numerical rank deficiency in matrices, by providing a column permutation as a part of QR factorizations [2]. The permutation is done so that most of the action is retained by the first vectors of the resulting orthogonal basis. Highly optimized implementations of algorithms for obtaining these factorizations have been proposed [1]. RRQR factorizations have been used as a part of some unsupervised feature selection algorithms. In [3], for instance, Boutsidis et al. propose to randomly construct several column subsets, permute them using an RRQR factorization and then choose the one that minimizes the reconstruction error. A deterministic algorithm for the same objective function is proposed in [4]. The authors do not employ RRQR factorizations, but exploit the nature of the problem to build a very efficient greedy selection of columns. A parallelized version of this approach can be found in [5], although it differs from our method in that it is designed for high dimensions instead of high row counts.

Other parallelized feature selection approaches have been proposed in recent years. In [6] a framework for choosing a column subset that is close to the best rank-k approximation is proposed. It relies on partitioning the data matrix

$A \in \mathbb{R}^{m \times n}$ into submatrices of at least k =rank(A) columns. In [7] a parallelized method based on mutual information is described, designed for discrete data. There also exist methods for supervised learning, such as [8], a MapReduce implementation of the minimum redundancy maximum relevance algorithm, [9] a parallelized feature selection algorithm for logistic regression and [10] a method for microarray data classification. In [11] an MPI-based algorithm for unsupervised feature selection is described, and in [12] a MapReduce implementation of a method based on rough set theory is presented. In [13] a parallelized implementation of an unsupervised feature selection method is proposed. Finally, a similar, more efficient parallelized algorithm is presented in [14].

3   Contribution

The approaches mentioned in Sect. 2 are different from the one presented in this paper in various aspects. In general, parallelized approaches present in the literature are designed for dealing with a large number of features instead of a large number of rows, designed for supervised learning or restricted to discrete data. The mentioned centralized methods could be adapted to big data sets by parallelizing some of the matrix operations involved, but high dimensionality could pose some challenges.

In this paper we present a proof for the preservation of the permutations yielded by rank-revealing QR factorizations under a certain transformation. This leads to an algorithm that can be used for unsupervised feature selection on big data sets. Our method presents the following desirable qualities: (1) it can deal with data sets of any number of rows, incurring only a linear increase in running time; (2) the form of the compact representation we use as input for the final factorization makes it possible to deal with high dimensionality, often with small or no loss in precision.

We perform experiments to validate the usefulness of our algorithm.

4   Permutation-Preserving Transformations for RRQR Factorizations

4.1   Notation

- $A_{i:}$ is the i-th row of matrix A ($A_{:i}$ is analogous for columns).
- $A_{i:j,k:l}$ is the submatrix of A that contains the rows and columns in the ranges i to j and k to l respectively.
- $A_{i:j,:}$ is the submatrix of A that contains rows i to j and all columns.
- Lowercase bold letters (e.g. $\boldsymbol{d}$) represent vectors.

In this section we prove that a big data set can be transformed into a compact form in order to enable the use of RRQR factorizations. Specifically, given a matrix $A \in \mathbb{R}^{m \times n}$, with rank($A$) = n, $m \geq n$ we can work with a compact matrix $\tilde{A} \in \mathbb{R}^{n \times n}$ and obtain the same result. This proves very useful in big data scenarios, where the value of m can grow indefinitely but the value of n (i.e. the number of features of our data set) usually stays the same or varies only slightly.

In order to facilitate the representation of some of the employed operations we define the following function.

Definition 1. *Given $m, n, i, j \in \mathbb{N}$, $1 \le i \le n$, $1 \le j \le n$ we define the following function.*

$$\phi_{ij} : \mathbb{R}^{m \times n} \to \mathbb{R}^{m \times n}$$

$$A \mapsto \phi_{ij}(A) \text{ such that } \begin{cases} \phi_{ij}(A)_{:i} = A_{:j} \\ \phi_{ij}(A)_{:j} = A_{:i} \\ \phi_{ij}(A)_{:k} = A_{:k} & \text{if } k \neq i, j \end{cases}$$

This function simply interchanges the columns i and j in matrix A.

Algorithm 1 is a pseudo-code representation of the traditional method for computing QR factorizations with column pivoting. In this version of the code we only pay attention to the resulting permutation, and insert zeros in the input matrix to update the column norms. The procedure can therefore be slightly different from other examples present in the literature. This algorithm functions in a simple fashion. At each iteration, the column with the largest norm is permuted to the front. Then, a Householder matrix is sought such that said column ends up being aligned with a vector of the canonical basis. The coefficient corresponding to that vector is then subtracted from all the columns.

Algorithm 1. Traditional algorithm for QR with column pivoting (permutation only)

1: **procedure** RRQR($A \in \mathbb{R}^{m \times n}$)

2:    $\boldsymbol{\pi} \leftarrow 1, \ldots, n$ ▷ Permutation vector

3:    **for** $i = 1, \ldots, n$ **do**

4:        $p \leftarrow \operatorname{argmax}_j \|A_{:j}\|_2, i \le j \le n$

5:        switch($\boldsymbol{\pi}, i, p$); switch(A, i, p)

6:        $A_{i:m, i+1:n} \leftarrow H A_{i:m, i+1:n}$ ▷ $H A_{i:m, i} = \pm \|A_{i:m, 1}\|_2 \mathbf{e}_1$

7:        $A_{i:} \leftarrow \mathbf{0}^T$

8:    **end for**

9:    Output $\boldsymbol{\pi}$

10: **end procedure**

We now proceed to prove the result introduced at the beginning of this section. Throughout this section we consider a data matrix $A \in \mathbb{R}^{m \times n}$, $m \ge n$, rank$(A) = n$ and consider $A = U\Sigma V^T$ to be its "thin" singular value decomposition (i.e. $U \in \mathbb{R}^{m \times n}, \Sigma \in \mathbb{R}^{n \times n}, V \in \mathbb{R}^{n \times n}$). We also consider a matrix $\tilde{A} = \Sigma V^T$, and use the tilde character to refer to variables of the algorithm running with $\tilde{A}$ as input (e.g. if π is the permutation variable for A, then $\tilde{\pi}$ is the permutation variable for $\tilde{A}$). The following lemma will be helpful at some steps of the proof.

Lemma 1. *For $i = 1, \ldots, n$, $\|A_{:i}\|_2^2 = \|\tilde{A}_{:i}\|_2^2$*

Proof. We have that $\Sigma V^T = U^T A$ and the n columns of U are pairwise orthogonal and span the column space of A. Therefore, the lemma holds.   □

Lemma 2. *Let us assume that at the beginning of iteration i of Algorithm 1, $\pi = \tilde{\pi}$ and $\|A_{:j}\|_2^2 = \|\tilde{A}_{:j}\|_2^2$ for $j = 1, \ldots, n$. Then, after iteration $i + 1$, $\pi = \tilde{\pi}$, $\|A_{:j}\|_2^2 = \|\tilde{A}_{:j}\|_2^2$ for $j = 1, \ldots, n$.*

Proof. At iteration i, let $p = \text{argmax}_j \|A_{:j}\|_2^2$, $\tilde{p} = \text{argmax}_j \|\tilde{A}_{:j}\|_2^2$ (line 4 of Algorithm 1). Since by assumption $\|A_{:j}\|_2^2 = \|\tilde{A}_{:j}\|_2^2$, then $p = \tilde{p}$.

Let $x = A_{i:m,p}$, $\tilde{x} = \tilde{A}_{i:n,p}$. At iteration i, Algorithm 1 builds $H \in \mathbb{R}^{m-i+1 \times m-i+1}$, $\tilde{H} \in \mathbb{R}^{n-i+1 \times n-i+1}$ so that

$$Hx = \pm\|x\|_2 e_1$$

$$\tilde{H}\tilde{x} = \pm\|\tilde{x}\|_2 \tilde{e}_1$$

where e_1 is the first vector of the canonical basis for $\mathbb{R}^{m-i+1}$, $\tilde{e}_1$ respectively for $\mathbb{R}^{n-i+1}$.

Let $B = H\phi_{0,p-i}(A_{i:m,i:n})$, $\tilde{B} = \tilde{H}\phi_{0,\tilde{p}-i}(\tilde{A}_{i:n,i:n})$, i.e., the input matrices after interchanging the columns and applying the Householder transformation. At line 7, the row i of matrix A is transformed into a zero row vector. Therefore, it is easy to see that after said line has been executed, for $j = i, \ldots, n$

$$A_{i:m,j} = B_{i:m,j} - B_{i,j} e_1$$

$$\tilde{A}_{i:n,j} = \tilde{B}_{i:n,j} - \tilde{B}_{i,j} e_1$$

Which means that the respective squared norms for $j = i, \ldots, n$ are now (note that all the entries above row i are zeroes)

$$\|A_{:j}\|_2^2 = \|B_{:j}\|_2^2 - B_{i,j}^2 \tag{1}$$

$$\|\tilde{A}_{:j}\|_2^2 = \|\tilde{B}_{:j}\|_2^2 - \tilde{B}_{i,j}^2 \tag{2}$$

Now, since H and $\tilde{H}$ are orthogonal and $p = \tilde{p}$, for $j = i, \ldots, n$ we have that

$$\|x\|_2^2 B_{i,j} = x^T B_{:j} = \tilde{x}^T \tilde{B}_{:j} = \|\tilde{x}\|_2^2 \tilde{B}_{i,j}$$

which implies that $B_{i,j} = \tilde{B}_{i,j}$. Since by orthogonality of H and $\tilde{H}$, $\|B_{:j}\|_2^2 = \|\tilde{B}_{:j}\|_2^2$, by Eqs. 1 and 2 $\|A_{:j}\|_2^2 = \|\tilde{A}_{:j}\|_2^2$.   □

Theorem 1. *Let $A \in \mathbb{R}^{m \times n}$, $\mathrm{rank}(A) = n$ and let $A = U\Sigma V^T$ be its "thin" singular value decomposition. If $\tilde{A} = \Sigma V^T$, then the permutation yielded by Algorithm 1 on matrices A and $\tilde{A}$ is the same.*

Proof. Since on entry the permutation vector is sorted in ascending order and by Lemma 1 the column norms are equal for both matrices, by Lemma 2 the permutation vector is the same at the end of all iterations. ∎

Theorem 1 allows for the design of a simple and efficient algorithm to rank and therefore select features from a big data set which. We have dubbed it Feature Ranking and Selection (FRANKS). Pseudo-code is offered in Algorithm 2. The main advantage of this algorithm is that there exist distributed algorithms to compute the singular value decomposition exactly, which is not the case of rank-revealing QR factorizations. Also, the form of matrix $\tilde{A}$ offers an additional advantage. Since $\tilde{A} = \Sigma V^T$, all rows below the numerical rank of the matrix can be discarded with no impact on the obtained permutation. Furthermore, in most data sets the last singular values are small, albeit non-zero. Therefore, some additional rows of $\tilde{A}$ can be discarded at a low cost in precision if the value of n and the rank of the matrix are large.

Algorithm 2. The FRANKS algorithm.

1: **procedure** FRANKS($A \in \mathbb{R}^{m \times n}$)

2:     Compute the SVD of A: $A = U\Sigma V^T$

3:     $\pi \leftarrow \mathrm{RRQR}(\Sigma V^T)$

4:     Output π

5: **end procedure**

---

5 Experimental Results

In this section we present experimental results to demonstrate the benefits of using our algorithm as a preprocessing step for performing machine learning tasks on large data sets. We run the k-means clustering algorithm on the widely used MNIST data set. MNIST is a set of 60,000 28×28 images of handwritten digits. Each image is represented as a 784-dimensional vector of brightness levels.

We run Algorithm 2 on this data set to obtain a feature ranking. Afterwards, we choose increasingly large subsets of features according to this ranking and run the k-means algorithm using only the chosen features. The quality of the clustering results is measured using the normalized mutual information (NMI) metric. In order to assess the effectiveness of our method we also run k-means using random subsets of features. We perform 25 runs and compute the average of the result, both for the random subsets and for the features chosen by our method. Figure 1 shows the NMI obtained in this experiment using increasingly large subsets of features. The result using all features (also averaged over 25 runs) is also shown for comparison. It can be clearly seen that our method provides small feature subsets (≈ 100) that can attain the same or even a better result than the whole feature subset. Figure 2 shows the running time of one k-means iteration using increasingly large subsets of features compared to the running time using all features. It can be seen that using over 100 features, where the NMI results equate or surpass the baseline result, the savings in running time are significant (Table 1).

Table 1. NMI for the k-means algorithm using the features selected by FRANKS and a random subset of features.

k	FRANKS	Random subset	k	FRANKS	Random subset
10	0.283039	0.155082	110	0.495866	0.348588
20	0.400929	0.199653	120	0.502660	0.353364
30	0.456524	0.240830	130	0.513387	0.378147
40	0.473640	0.269519	140	0.508007	0.373178
50	0.484083	0.264572	150	0.507678	0.381401
60	0.464566	0.275641	160	0.498707	0.402576
70	0.474112	0.302368	170	0.499427	0.393623
80	0.483499	0.318010	180	0.505391	0.400914
90	0.489321	0.333131	190	0.508073	0.410332
100	0.483821	0.347894	200	0.494279	0.412137

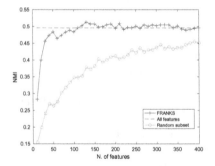

Fig. 1. NMI for the k-means algorithm using the features selected by FRANKS and a random subset of features, compared to the result using the whole feature subset.

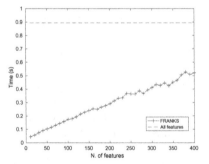

Fig. 2. Running time of one k-means iteration using different numbers of features, compared with the time with all features.

6   Conclusions and Future Work

We have presented a proof of the permutation-preserving nature of orthogonal transformations for rank-revealing QR factorizations. Leveraging this result we have developed an efficient algorithm for deterministic unsupervised feature selection on big data sets. Our method can deal with data sets of any number of rows, and high dimensionality can often be addressed with small or no loss in precision. Experimental results show that this method can yield benefits both in clustering results and running time using the k-means algorithm on the MNIST data set. In the future, we plan to extend the algorithm to decrease the loss when dealing with high dimensional data sets, as well as work on methods for improving the loss in linear approximation using the chosen features as a basis.

Acknowledgements. The research leading to these results has received funding from the European Union under the FP7 grant agreement n. 619633 (project ONTIC) and H2020 grant agreement n. 671625 (project CogNet).

References

1. Quintana-Ortí, G., Sun, X., Bischof, C.H.: A blas-3 version of the qr factorization with column pivoting. SIAM J. Sci. Comput. **19**(5), 1486–1494 (1998)
2. Chan, T.F.: Rank revealing qr factorizations. Linear Algebra Appl. **88**, 67–82 (1987)
3. Boutsidis, C., Mahoney, M.W., Drineas, P.: Unsupervised feature selection for principal components analysis. In: Proceedings of the 14th ACM SIGKDD International Conference on Knowledge Discovery and Data Mining, pp. 61–69. ACM (2008)
4. Farahat, A.K., Ghodsi, A., Kamel, M.S.: An efficient greedy method for unsupervised feature selection. In: 2011 IEEE 11th International Conference on Data Mining (ICDM), pp. 161–170. IEEE (2011)
5. Farahat, A.K., Elgohary, A., Ghodsi, A., Kamel, M.S.: Distributed column subset selection on mapreduce. In: 2013 IEEE 13th International Conference on Data Mining (ICDM), pp. 171–180. IEEE (2013)
6. Pi, Y., Peng, H., Zhou, S., Zhang, Z.: A scalable approach to column-based low-rank matrix approximation. In: Proceedings of the Twenty-Third International Joint Conference on Artificial Intelligence, pp. 1600–1606. AAAI Press (2013)
7. Sun, Z., Li, Z.: Data intensive parallel feature selection method study. In: 2014 International Joint Conference on Neural Networks (IJCNN), pp. 2256–2262. IEEE (2014)
8. Reggiani, C., Le Borgne, Y.-A., Pozzolo, A.D., Olsen, C., Bontempi, G.: Minimum redundancy maximum relevance: Mapreduce implementation using apache hadoop. In: BENELEARN 2014, p. 2 (2014)
9. Singh, S., Kubica, J., Larsen, S., Sorokina, D.: Parallel large scale feature selection for logistic regression. In: SDM, pp. 1172–1183. SIAM (2009)
10. Bolón-Canedo, V., Sánchez-Maroño, N., Alonso-Betanzos, A.: Distributed feature selection: An application to microarray data classification. Appl. Soft Comput. **30**, 136–150 (2015)

11. Zhao, Z., Zhang, R., Cox, J., Duling, D., Sarle, W.: Massively parallel feature selection: an approach based on variance preservation. Mach. Learn. **92**(1), 195–220 (2013)
12. He, Q., Cheng, X., Zhuang, F., Shi, Z.: Parallel feature selection using positive approximation based on mapreduce. In: 2014 11th International Conference on Fuzzy Systems and Knowledge Discovery (FSKD), pp. 397–402. IEEE (2014)
13. Ordozgoiti, B., Gómez Canaval, S., Mozo, A.: Massively parallel unsupervised feature selection on spark. In: Morzy, T., Valduriez, P., Bellatreche, L. (eds.) ADBIS 2015. CCIS, vol. 539, pp. 186–196. Springer, Heidelberg (2015)
14. Ordozgoiti, B., Canaval, S.G., Mozo, A.: Parallelized unsupervised feature selection for large-scale network traffic analysis. Proc. ESANN **2016**, 617–622 (2016)

BAC: A Bagged Associative Classifier for Big Data Frameworks

Luca Venturini$^{(\boxtimes)}$, Paolo Garza, and Daniele Apiletti

Dipartimento di Automatica e Informatica,
Politecnico di Torino, Corso Duca degli Abruzzi 24, Torino, Italy
{luca.venturini,paolo.garza,daniele.apiletti}@polito.it

Abstract. Big Data frameworks allow powerful distributed computations extending the results achievable on a single machine. In this work, we present a novel distributed associative classifier, named BAC, based on ensemble techniques. Ensembles are a popular approach that builds several models on different subsets of the original dataset, eventually voting to provide a unique classification outcome. Experiments on Apache Spark and preliminary results showed the capability of the proposed ensemble classifier to obtain a quality comparable with the single-machine version on popular real-world datasets, and overcome their scalability limits on large synthetic datasets.

Keywords: Associative classifiers · Bagging · Machine learning · MapReduce · Apache spark

1 Introduction

Big Data frameworks are becoming increasingly popular thanks to their technological maturity, the wide-spread availability, and the value they allow to extract from very large data sets. Even if *volume, velocity,* and *variety* are the typical characteristics of Big Data [1], others are often considered critical for businesses: among the top Vs of Big Data, *value* is probably the most valuable. Hence, Big Data frameworks are technological enablers to not only process larger and more complex data sets, but also to extract higher value from any dataset.

In this work, we aim at distributing the training phase of a special kind of classifiers, namely associative classifiers, whose training is highly sequential in its very own nature. The distribution of the work on a cluster of machines could have several advantages, among which a performance boost, or an increase of the dataset size which can be analyzed. Preliminary experimental results exploiting a distributed approach on Apache Spark applied to real-world and synthetic datasets showed promising results: ensembles of multiple models trained on small subsets of the original dataset (as small as 10 %) lead to accuracies comparable with the single model trained on the whole dataset. Our method thus proves to be a viable solution to distribute the workload of this task, without compromising on the quality of the results.

© Springer International Publishing Switzerland 2016
M. Ivanović et al. (Eds.): ADBIS 2016, CCIS 637, pp. 137–146, 2016.
DOI: 10.1007/978-3-319-44066-8_15

This paper is organized as follows. Section 2 describes the data mining background, Sect. 3 presents the proposed approach, Sect. 4 discusses the experimental results, and Sect. 5 compares our approach with previous works. Finally, Sect. 6 draws conclusions and presents future developments.

2   Problem Statement

In this section, we state the classification problem and describe how association rules are usually exploited to solve it. Finally, we describe the centralized L^3 associative classifier [2], since the distributed BAC algorithm proposed in this paper is based on L^3.

2.1   Classification Problem

The input dataset $\mathcal{D}$ is represented as a relation R, whose schema is given by k distinct attributes $A_1 \ldots A_k$ and a class attribute C. Each record in R can be described as a collection of pairs *(attribute, integer value)*, plus a class label (a value belonging to the domain of class attribute C). Each pair *(attribute, integer value)* will be called *item* in the following. A training record is a tuple in R where the class label is known, while an unlabelled record is a tuple in R where the class label is unknown. The attributes can be either categorical or continuous attributes. For categorical attributes, for each attribute, all values in its domain are mapped to consecutive positive integers. In the case of continuous attributes, the values of each attribute are first discretized into intervals, and then intervals are mapped into consecutive positive integers.

A classifier is a function from $A_1, \ldots, A_n$ to $\mathcal{C}$, that allows the assignment of a class label to an unlabelled records. Given a collection of training data, the classification task is the generation of a classifier able to predict the class label for the unlabeled data with high accuracy.

2.2   Associative Classification

Associative classifiers (e.g., [2,3]) are well-known classifiers based on association rules [4]. Association rules [4] are rules in the form $X \rightarrow Y$, where both X and Y are set of items. When using them for classification purposes, X is a set of items (i.e., (attribute, value) pairs), while Y is a class label. A record d is said to match a collection of items X when $X \subseteq d$. The quality of an association rule is measured by two parameters, its support, given by the number of records matching $X \cup Y$ over the number of records in the training dataset, and its confidence given by the number of records matching $X \cup Y$ over the number of records matching X. Hence, the classification task can be reduced to the generation of the most appropriate set of association rules for the classifier.

The model generation phase of the associative classifiers (e.g., [2,3]) is based on two steps: (i) Extraction of all the classification rules with a support higher than a minimum support threshold *minsup* and a minimum confidence threshold

$minconf$ and (ii) Rule selection by means of the database coverage technique. Step (i) extracts all the rules that are potentially significant, in terms of frequency and confidence, while the Step (ii) selects the best rules by evaluating their accuracy.

Once the model is built, the prediction of the class label of a generic unlabelled record d is performed by considering the the first rule r matching d, based on a quality ranking, and the class label of r is used to label d.

The L^3 classifier. The proposed distributed algorithm (Sect. 3) is based on the L^3 associative classifier. Hence, we describes the centralized version of L^3. The model generation phase of L^3 algorithm is based on two main steps: (i) frequent classification rule mining and (ii) rule selection by means of a lazy database coverage technique. To address the mining step efficiently, L^3 exploits an FP-growth like association rule mining algorithm. The exploited mining algorithm is optimized to directly extract classification rules (i.e., association rules with a class level as consequent).

Once the potentially large of set of frequent classification rules is available, a lazy pruning step, based on the database coverage approach is applied. Before performing the pruning phase, a global order is imposed on the extracted frequent classification rules. Let r_1 and r_2 be two classification rules. Then r_1 precedes r_2, denoted as $r_1 > r_2$ if

1. $\mathrm{conf}(r_1) > \mathrm{conf}(r_2)$, or
2. $\mathrm{conf}(r_1) = \mathrm{conf}(r_2)$ and $\sup(r_1) > \sup(r_2)$, or
3. $\mathrm{conf}(r_1) = \mathrm{conf}(r_2)$ and $\sup(r_1) = \sup(r_2)$ and $\mathrm{len}(r_1) > \mathrm{len}(r_2)$, or
4. $\mathrm{conf}(r_1) = \mathrm{conf}(r_2)$ and $\sup(r_1) = \sup(r_2)$ and $\mathrm{len}(r_1) = \mathrm{len}(r_2)$ and $\mathrm{lex}(r_1) > \mathrm{lex}(r_2)$

where $\mathrm{len}(r)$ denotes the number of items in the body of r, and $\mathrm{lex}(r)$ denotes the position of r in the lexicographic order on items.

After the rule sorting operation, only the rules satisfying the chi-square test, at a significance level of 95 % are selected (i.e., the rules with a correlation between the antecedent and the consequent of the rule).

Finally, the lazy pruning approach is applied. The idea behind the lazy pruning is to discard only the rules that do not correctly classify any training record, i.e., the rules that only negatively contribute to the classification of training records and organize the other rules in two levels (first level and second level). To achieve this goal, the lazy pruning approach splits the frequent association rule set in three subsets:

- **First level rules**. This rule set contains the mining set of rules that is needed to properly "cover" the training data. This set of rules represents the main characteristics of the majority of the training records.
- **Second level rules**. These rules are not included in the first level, but are potentially useful to represent the characteristics of "special" records that are slightly different with respect to the most common ones appearing in the training set. These rules are also called "spare" rules and are useful to properly label new unlabelled data not represented by the rules of the first level.

– **Harmful rules**. Some rules, even when applied on the training set, always perform wrong predictions. These rules must be removed since they contribute only negatively to the quality of the generated model.

To identify the three described subsets, L^3 applies a database coverage technique. Specifically, L^3 considers one rule r at a time in the sort order and selects the training records matched by r. For each matched record d, L^3 checks also if r classifies properly d. If r classifies properly at least one training record, then r is stored in the first level of the classifier. Differently, if all the training records matched by r have a class label different from the one of r, then r is discarded (i.e., it is an harmful rule). If r does not match any training data then r is store in the second level of the final classifier. Once r has been analysed, all the training records matched by r are removed from the training set and the next classification rule is analysed by considered only the remaining training records.

To predict the class label of an unlabelled record d, the rules of the first level are initially considered. If no rule in the first level matches d, the second level is used.

2.3   Bagging

The bagging technique is frequently used to build accurate models by combining a set of "weak" classifiers. The basic idea is that a set of classifiers can provide better predictions than a single model.

The bagging technique works as follows.

1. Generate N datasets by applying random sampling with replacement on the training dataset.
2. Build N classification models (one for each dataset generated during Step 1).
3. Predict the class label of the new unlabelled records by using a majority voting approach combining the predictions of the N classifiers that have been built during Step 2.

3   BAC: Bagged Associative Classifier

This Section describes the proposed approach to scale on different machines the training of an associative classifier, namely L^3, which is presented in Sect. 2. The Big Data framework we exploited is the well-known Apache Spark, which poses some specific technological issues to the design of an associative classifier.

We recall, as mentioned in Sect. 2, that an important phase of the generation of such a classifier is represented by the database coverage (lazy pruning in L^3). Unfortunately, the database coverage algorithm does not fit a map-reduce approach, as it is sequential in its nature: the database must be covered in the strict order of the rules for the algorithm to be effective. Such requirement prevents most of the work of this phase to be executed in parallel.

To cope with the scalability of the process, in order to fully exploit the distributed framework, we adopt bagging. The solution consists in generating several models, each one from a portion of the original full dataset. Each model can then be trained independently, thus also in parallel on multiple machines. The model trained locally on each portion of the dataset is a variant of L^3, where we discharge the second level rules after the database coverage phase. The FP-growth algorithm used is the one implemented in Apache Spark, slightly modified to run locally. The ensemble of the single models eventually generate a single prediction by majority voting.

In the following we provide details about the split of the dataset among the different machines (i.e., Apache Spark workers). The ensemble of models is eventually collected and can be used on any number of machines to classify new records.

3.1   Dataset Distribution

Each model, as said, is trained on a different portion of the original dataset. Each portion is drawn by sampling the original records with replacement, as this method is well-proven in literature, like mentioned in Sect. 2.3. Moreover, whereas this method allows for obtaining any sort of combination for number of models and partitions size, sampling without replacement would limit the total number of records to the original dataset size. In other words, sampling with replacement can produce, for example, multiple models trained on a dataset as large as the original dataset (and still different), or even larger; or we might have a dozen models, each trained on half of the original dataset, while without replacement we would have just two halves available.

The core APIs of Apache Spark provide a method for performing sampling with replacement. The method is characterized by two parameters: the input dataset D and a real number λ that is used to specify the size of the sample as a fraction of the input dataset. Specifically, a sample of size $|D| \times \lambda$ is generated. The provided method performs a sampling with replacement, generating k copies for each input record, where k is drawn from a Poisson distribution.

In our case, we are interested in N samples, each one with a size equal to $|D| \cdot f$, since we want to build N models for N different samples. f is real parameter of the algorithm and it is used to specify the size of each sample. We can generate the N samples invoking, sequentially, the Spark sampling API N times setting λ to f. However, this approach will reduce the parallelism of our algorithm. Hence, we decided to use another approach that allows us generating simultaneously N samples, each of size $|D| \cdot f$. Specifically, we proceed in the following way. First, we sample the dataset D with $\lambda = N \cdot f$, as to have an expected single sample with a total size equal to $|D| \cdot N \cdot f$. Now we need to split this sample in N subsamples. We can perform this operation by using the parallel Spark API `repartition` that splits the input data in a user-specified number of partitions (in our case we set it to N). Hence, each partition contains a sample of size $|D| \cdot f$ of the input dataset.

Once obtained N samples with the desired fraction of data, each on a separate partition, we can simply call `mapPartitions` in Spark to apply the same function on each, in our case the training function of the model.

4   Experimental Validation

To validate the proposed approach, we implemented BAC in Scala on top of Apache Spark. Experiments aim at assessing the usefulness of a distributed app-roach against two alternatives: (i) working on the whole dataset on a single machine, which is not always feasible, and (ii) working on a sample of the origi-nal dataset, that is always feasible for small portions, but at the costo of lower quality of the model (e.g., accuracy). Experiments also evaluate the effect of the number of estimators (models) of the ensemble on the final quality of the whole ensemble model.

As evaluation criterion, we focus primarily on the accuracy, computed on a 10-fold cross-validation of the selected datasets. For each mean accuracy, we computed a 95 % confidence interval based on the standard error of the mean, using a t-student statistics. We also computed the average time for training.

Three very popular datasets have been used for the experiments: yeast, nursery, and census, from the UCI repository [5]. We have chosen these datasets as they are heterogeneous in dimensions, shape and distribution, and state-of-art associative classifiers do not perform well, so that there is still margin for improvements. The continuous attributes have been discretized applying the entropy-based discretization technique [6]. Since the bigger of the three datasets counts for 30162 records only, we generated a fourth synthetic dataset containing 1 million tuples and 9 attributes with different distributions, using the IBM data generator. Continuous attributes have then been discretized to 10 bins each.

All experiments share the same minimum support threshold (1 %) and the same minimum confidence value (50 %), as in previous works of associative classi-fiers [2] they proved to generate a good amount of significant rules. Experiments were performed on a cluster with 30 worker nodes running Cloudera Distribution of Apache Hadoop (CDH5.5.1), which comes with Spark 1.5.0. The cluster has 2.5 TB of RAM, 324 cores, and 773 TB of secondary memory. The size of each container has been set to 2 GB.

4.1   Results

Focusing on real datasets, that are yeast, nursery, and census, Fig. 1(a), (b) and (c) show the average accuracy. The dashed line represents the accuracy obtained by the classifier trained over the whole dataset on a single machine. The range of its confidence interval, highlighted in the figures, serves as reference for our evaluation. On the x axes we have the number of models trained, each on a 10 %-portion of the original dataset. For $x = 1$, we have the simple sampling. For yeast (Fig. 1(a)), we see how the simple sampling can reduce the total accuracy of more than 8 % with respect to the level of the classifier trained on the whole dataset. In nursery (Fig. 1(b)) we see the same behaviour. Here the drop is less marked, less than 2 %, but still significant if we compare the confidence intervals. Curiously, census (Fig. 1(c)) shows a completely different behaviour. In this dataset, representing a US census of the population, sampling obtains an accuracy even higher than our reference. This surprising result can be due to a simpler, even if weaker, model less prone to overfitting the training data. The 10-fold cross-validation indeed penalizes overfitting models.

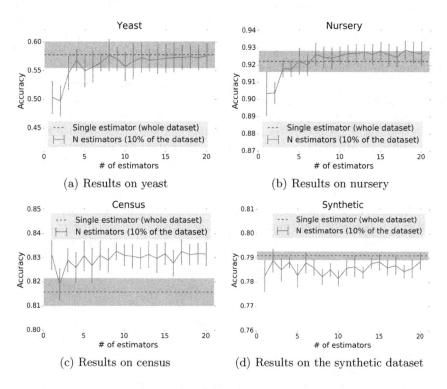

(a) Results on yeast

(b) Results on nursery

(c) Results on census

(d) Results on the synthetic dataset

Fig. 1. Accuracy results

Let us now inspect the results of BAC, with an increasing number of models, up to 20. For all datasets, the accuracy level stabilizes way before 10 models. We see that yeast (Fig. 1(a)), for example, enters the confidence interval of the reference model from 3 models on, becoming statistically indifferent. Large confidence intervals are a peculiarity of the k-fold cross-validation, and highly depend on the dataset. It is as well a measure of the robustness of the model towards different training sets. Nursery (Fig. 1(b)) and census (Fig. 1(c)) confirm the same trend as yeast, with BAC entering the confidence interval of the centralized classifier from 3 models onward. These results are interesting, since 3 models cover less than a third of the original dataset, and they are enough to reach similar accuracies to our reference model. This means also that distributing the dataset only once, i.e., 10 models of 10 % each, we can be confident enough of having entered already the steady region, without the need of more replicas/models. Furthermore, having a number of models equal to one corresponds to the simple sampling, whereas two models would affect the contribution of the majority voting among models. Thus 3 is also the minimum sensible setting of this parameter.

Results for the synthetic dataset are shown in Fig. 1(d). The dataset counts for a million records, and tries to emulate a possible use case of BAC, to show off its potential on large dataset. Firstly, the accuracy of the reference classifier is characterized by a very narrow confidence interval. This, as mentioned above, comes from the shape and distribution of the dataset: since it is synthetic, the distribution of the attributes and of the labels is uniform among the folds, thus the standard error of the mean accuracy is very little compared to the real datasets, feature that is further sharpened by the greater size of the folds. Sampling the 10 % of the dataset results in an almost negligible detriment to the quality, less than a point of accuracy. This outcome is not surprising, as the distribution of the attributes in the sample can not vary too much from the whole, being the attributes generated from a given distribution. The accuracy of BAC increasing the number of models steadily stays in the range of the single sample, often overlapped with the results of the reference model, showing that adding data from the remainder of the dataset does not add information to our model. Though in this case we cannot conclude for absolute better performances on either side (namely, BAC or single-machine), we see an interesting fact in the way these results were obtained. On the cluster described at the beginning of the section, indeed, the generation of the reference classifier, the single model trained on the whole dataset, failed, running out of the memory available to its (single) container. To obtain the value of reference plotted in Fig. 1(d) we executed the training on a standalone Spark machine, where we set the amount of memory for the Java VM to the whole RAM available on the machine itself (32 GB). All experiments for BAC, instead, completed successfully, proving that distributing workload can be a way to overcome the limits of the single machine and expand the size of the explorable datasets.

Finally, Table 1 shows the average training time for the 10 % sample, for BAC with 10 models on a tenth of the data each, and the reference model, that is the classifier trained on the whole data. As said, the last-mentioned classifier failed its execution for the synthetic dataset, so its timing is not fairly comparable with the others. On the three real datasets, we notice that, unsurprisingly, simple sampling outperforms the more accurate models. BAC is the slowest, as the overhead of the distributed framework and the communication costs are very high for such small datasets. On the larger synthetic dataset, these overheads are absorbed by the real computational costs, resulting in little difference between one or 10 machines working on models of the same size/complexity. We need further investigations to show the real impact of parallelization on a real dataset of this size or larger.

To sum up, from these results we can conclude that:

1. the mere sampling is not always sufficient to reach a good accuracy,
2. training an associative classifier over the whole dataset is not always feasible,
3. bagging is a viable solution to reach the quality of a single classifier trained on the whole dataset, and offers an easy way to distribute the work among multiple workers.

Table 1. Average training time of the different approaches.

Dataset	Records	Avg training time for a fold [ms]		
		1 model, 10 %	10 model, 10 %	1 model, 100 %
Census	30162	57692	340020	302147
Nursery	12960	983	3457	2181
Yeast	1484	642	1799	850
Synthetic	1000000	14235	15150	n.a

5   Related Work

The classification problem is a well-known problem and many approaches have been proposed to solve this problem [3,7,8] Among the others, associative classification is a well-known technique for structured data classification [2,3]. Associative classifiers are accurate. However, their are based on two complex and time consuming steps: (i) association rule mining and (ii) database coverage [2]. Out paper is the first to propose an associative classifier based on Spark to addresss the complexity of the two steps by means of a distributed approach.

The increase of the overall accuracy of the predictions is also addressed by means of ensemble techniques [6]. The paper [9] analysed the impact of the boosting ensemble technique when a set of associative classifiers are used as building block. However, the impact of the bagging ensemble approach based on associative classifiers has never been analysed. Differently from [9], in this paper we perform this analysis by proposing a bagging version of an associative classifier.

6   Conclusions and Future Work

In this paper, a novel distributed associative classifier based on bagging has been proposed. Preliminary experiments showed that bagging can achieve the quality of the single-machine version, with an accuracy always better or as good as the sampling-only approach. Therefore, BAC proved to be a simple and effective way to distribute the workload among the machines of a cluster. The achieved results are promising and hence we are performing further analysis, on other real-datasets, to analyse in more detail both the quality, in terms of accuracy, and the execution time of BAC. We also aim at improving BAC by evaluating a unified rule-generation phase, shared among the machines, to further reduce the total memory used by the cluster. Additional investigations should also identify the limits of this approach, such as dataset cardinality, dimensionality, and density.

Acknowledgment. The research leading to these results has received funding from the European Union under the FP7 Grant Agreement n. 619633 ("ONTIC" Project).

References

1. Tsai, C.W., Lai, C.F., Chao, H.C., Vasilakos, A.V.: Big data analytics: a survey. J. Big Data **2**(1), 1–32 (2015)
2. Baralis, E., Garza, P.: A lazy approach to pruning classification rules. In: ICDM 2002, Maebashi, Japan, December 2002
3. Liu, B., Hsu, W., Ma, Y.: Integrating classification and association rule mining. In: KDD 1998, New York, NY., August 1998
4. Agrawal, R., Imilienski, T., Swami, A.: Mining association rules between sets of items in large databases. In: SIGMOD 1993, Washington DC., May 1993
5. Blake, C., Merz, C.: UCI repository of machine learning databases (1998)
6. Han, J.: Data Mining: Concepts and Techniques. Morgan Kaufmann Publishers Inc., San Francisco, CA, USA (2005)
7. Quinlan, J.: C4.5: program for classification learning. Morgan Kaufmann, San Mateo (1992)
8. Rokach, L., Maimon, O.: Data Mining with Decision Trees: Theory and Applications. World Scientific Publishing Co. Inc., River Edge, NJ, USA (2008)
9. Sun, Y., Wang, Y., Wong, A.K.C.: Boosting an associative classifier. IEEE Trans. Knowl. Data Eng. **18**(7), 988–992 (2006)

Spark2Fires: A New Parallel Approximate Subspace Clustering Algorithm

Bo Zhu[✉] and Alberto Mozo

Universidad Politécnica de Madrid, Madrid, Spain
bozhumatias@ict-ontic.eu, a.mozo@upm.es

Abstract. Subspace clustering is an interesting investigation field that has been intensively studied in the last two decades. The objective of subspace clustering is to find all lower-dimensional clusters hidden in subspaces of high dimensional data. Although the majority of existing subspace clustering algorithms adopt certain heuristic pruning techniques to reduce the search space, the time complexity of such algorithms remain exponential with regard to the highest dimensionality of hidden subspace clusters. Even with help of parallelism, these techniques will require extremely high computational time in practice. In this paper we propose a novel subspace clustering technique that reduces the exponential time complexity to quadratic via approximation. We also provide a parallel implementation of proposed algorithm on top of Apache Spark to further accelerate our approach on large data sets. Preliminary experiment results show our algorithm performs much better especially considering the scalability with regard to the dimensionality of hidden clusters.

Keywords: Subspace clustering · Big data · Approximation

1 Introduction

Clustering techniques group similar points into entities called clusters. Usually the similarity between different points are measured by their pairwise distances using a predefined distance function. These similarities are used in the point assignment to the nearest centroid for partitioning-based methods, or in the nearest neighborhood query for density-based methods. However, as the dimensionality increases, distance measures fail to cluster objects well, because the existence of irrelevant dimensions will lead to distance distortions between points. Traditional clustering algorithms do not pay much attention to the fact that clusters can hide in different subspaces of the full feature space. For instance, in the field of bio-informatics, different groups of genes are relevant and only relevant to their corresponding symptom.

Dimensionality reduction techniques such as Principal Component Analysis are useful to transform the original large feature space to a new low-dimensional and informative space, but they scarify the interpretability of generated results. Feature selection techniques try to find a optimal subset of the whole feature set

© Springer International Publishing Switzerland 2016
M. Ivanović et al. (Eds.): ADBIS 2016, CCIS 637, pp. 147–154, 2016.
DOI: 10.1007/978-3-319-44066-8_16

by eliminating irrelevant or redundant dimensions. But these techniques usually select a superset of proper subspaces for each cluster. This will still lead to distance distortions and negatively affect the performance of subsequent clustering process. In addition, feature selection methods fail to find a proper subset for overlapping subspaces.

Since clusters are embedded in different subspaces of the original feature space, many subspace clustering algorithms have been proposed to find all the hidden lower-dimensional clusters in the context of high dimensional data clustering.

2   Related Work

The state of art algorithms can be roughly classified into bottom-up and top-down algorithms, and bottom-up algorithms are further divided into grid-based and density-based methods.

Top-down methods initial by taking samples from the whole data set considering all features, and iteratively discover and refine clusters with smaller subspaces. PROCLUS [1] is the first top-down algorithm that requires the average dimensionality of clusters as input parameter, which is difficult to determine beforehand. The use of sampling technique may cause PROCLUS to miss some entire clusters. DSKmeans [2] is a recent iterative subspace clustering algorithm which needs the number of clusters and a proper initialization set of centroids. SAP [3] maintains a feature weight vector for each point and iteratively updates this vector to finally generate proper subspaces for different clusters. The main drawback of SAP is that the iterative update of affinity matrix is computation-intensive and do not scale well to large data sets.

Grid-based methods partition data space into non-overlapping cells by discretizing each dimension into small units. A global density threshold is used in many grid and density based subspace clustering algorithms to determine whether a cell is dense. Finally adjacent dense cells in the same subspace are merged together to form clusters. CLIQUE [4] is recognized as the pioneering grid-based method for subspace clustering. An anti-monotonicity property is first used to prune sparse candidates. Mafia [5] improves CLIQUE by generating an adaptive grid in order to have less candidates in higher dimensions. Parallelism is also introduced to strengthen scalability and efficiency. Another variant of CLIQUE is ENCLUS [6], which uses fixed grid to search for interesting subspaces that may contain clusters based on entropy. SCHISM [7] extends CLIQUE by introducing variable density thresholds adapting to increasing dimensionality. nCluster [8] further improves CLIQUE by partitioning each dimension into many overlapping small bins, and then select only maximal nClusters for higher dimensionality.

Density-based methods intend to find dense regions that are separated by sparse ones by querying neighborhood points within a predefined distance threshold considering only relevant dimensions. SUBCLU [9] and its parallel variant CLUS [10] are typical examples of this category. INSCY [11] improves by introducing SCY-tree indexing technique and prune redundant clusters continuously.

DB-CSC [12] combines density-based subspace clustering with dense subgraph mining.

In several experimental evaluation works [13–15] some meaningful comparison results are provided: the efficiency and accuracy of existing grid-based and density-based subspace algorithms is highly dependent on the proper tuning of input parameters. It also has been observed in [13,16] that a global density threshold will cause a bias to a certain dimensionality. The quality of clustering result of some top-down algorithms is also dependent on the "guessing" of crucial parameters in [15].

In general, top-down approaches and density-based bottom-up approaches generally do not scale well w.r.t. data size. Grid-based subspace clustering algorithms that follow the bottom-up approach scale linearly with regard to the size of datasets, but they suffer from extremely highly time complexity, i.e. exponential to the dimensionality of hidden clusters.

In this paper we propose a novel subspace algorithm that significantly reduce the time complexity of such algorithms, from exponential to super-quadratic w.r.t. the dimensionality of hidden clusters. Preliminary experiment results show that this new algorithm has better efficiency and scalability.

3   Method

Grid-based (or cell-based) subspace clusters are defined in [14] using a global density threshold.

For each dimension i in a subspace S contained in the full dimension space D, the number of objects $|O_i|$ of a grid-based subspace cluster in intervals $I_i = [l_i, u_i] \subset V_i = [V_{min}, V_{max}]$ should be larger than a threshold τ, i.e.

$$\forall i \in S, O = \{o | o \in I_i\}, |O_i| >= \tau$$

Dimensionality bias is first discussed in [16] referring to the fact that a tight global density threshold may distinguish clusters well from noise in low dimensions but lose high dimensional clusters; while a loose one can detect clusters in high dimensions as well as extensive untranslatable low dimensional clusters.

To avoid the effect of dimensionality bias, we conduct subspace clustering based on the mutual similarities between all based cluster pairs to avoid the use of sensitive global parameters. Similar idea was proposed in [17] and we develop our algorithm improving from Fires and combining it with our previous grid-based clustering method PSCEG [18].

Our new algorithm, Spark2Fires, consists of three principal steps. It first starts from each of the original features, and partitions the value range of each feature into 10,000 equal-size small intervals and build the corresponding histogram. The frequency of each interval is further input to a weighted version of DBSCAN algorithm to accelerate the clustering process. The generated 1-dimensional clusters are saved as base clusters. The similarity between two base clusters is defined as the size of their intersection. In the second step, it first finds those large base clusters, the size of whose intersection with its most similar base cluster, is larger than $2/3 * averagesizeofallbaseclusters$, while at the same time the size of the complement set is also larger than such value. After splitting these base clusters into two smaller clusters, the set of base clusters is updated. Given K an input, we then find the K nearest neighbors (KNNs) for each base cluster. If the size of intersection of KNNs for two base clusters is larger than a given input μ, they are a best merge candidate for each other. If the size of best merge candidates set is larger than another given input minClu, such base cluster becomes a best merge cluster. Following these definitions, in order to group up similar base clusters, we first identify the set of best merge clusters. Then we find all pairs of best merge cluster who are mutual best merge candidates. Finally we group up all best merge cluster pairs that have one base cluster in common. In step 3, we transform the base cluster groups into hyperrectangles to formally define them as grid-based subspace clusters and get the final point assignments.

Algorithm 1. Spark2Fires

1: V_i : value domain of dimension i, N : total number of objects

Phase 1: GenerateBaseClusters

2: **for** each dimension $i \in S$ **do**
3:     sc.textFile: read in data file
4:     map: Populate objects into 10,000 discretized equal-sized bins of V_i
5:     reduceByKey: Get global frequencies of each bin
6:     mapValues: Assign the frequency to the lower bound of bins as its weight.
7:     mapValues: WeighedDBSCAN(epsilon $= \theta * V_i$, minPoints $= \theta * N$, listOfWeight)
8:     map: generateBaseClusters()
9: **end for**

Phase 2: GenerateBaseClusterGroups

10: map: findLargeCluster2split()
11: flatMap: split()
12: broadcast: baseClusterList.collect()
13: mapValues: calculateSimilarity()
14: mapValues: findKNearestNeighbors()
15: mapValues: findBestMergeCandidates()
16: mapValues: findBestMergeClusters()
17: reduce: findBestMergeClusterPairs()
18: mapValues: generateBaseClusterGroups()

Phase 3: GenerateFinalSubspaceClusters

19: map: transform2FinalSubspaceClusters()
20: reportClusters()

4   Preliminary Experimental Results

In this section we provide preliminary experimental results of our parallel algorithm considering accuracy and scalability with regard to the size and dimensionality of dataset, and the dimensionality of hidden subspace clusters. The preliminary experiments reported here were executed on a commodity cluster of 17 nodes. Each node consists of 4 Core(TM) i5 CPUs with 8 GB of RAM. Spark was configured to use one executor per core with 1 GB of memory each.

Synthetic datasets are used for the experiments in order to know the ground truth of the generated subspace clusters. We implemented a public data generator used in [3] to generate our datasets. Both scalability and accuracy of clustering results are reported below.

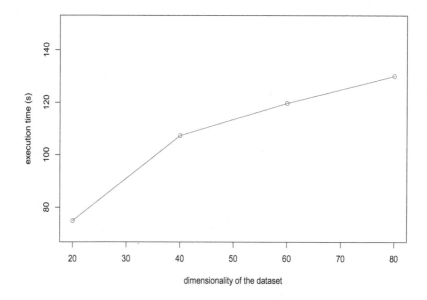

Fig. 1. Execution time w.r.t. dataset dimensionality

4.1   Scalability and Accuracy

To investigate the scalability with regard to dimensionality of datasets, we generated a series of test data with incrementing dimensionality containing 0.5 million points. In each data file there are four possibly overlapping 4-dimensional subspace clusters, of different sizes. We conducted Spark2Fires algorithm on these datasets and it successfully found all clusters in their correct subspaces despite of the increasing dimensionality of the dataset. Figure 1 shows that our algorithm scales linearly w.r.t. data dimensionality.

Figure 2 shows the scalability w.r.t. the dimensionality of hidden clusters. All the tested datasets have 20 dimensions and 1 million points in total, and

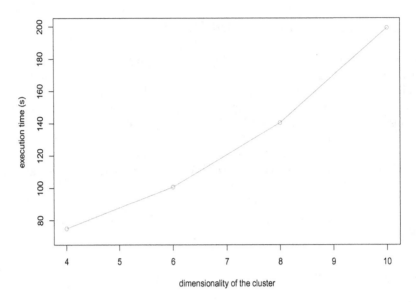

Fig. 2. Execution time w.r.t. cluster dimensionality

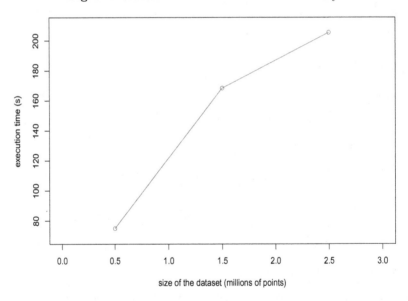

Fig. 3. Execution time w.r.t. size of datasets

4 possibly overlapping clusters that contain varying dimensionality, which hide in the full feature space. Our algorithm succeeded to find all four clusters correctly and efficiently without any sacrifice of accuracy. From Fig. 2 we can see that the execution time grows quadratically with increasing dimensionality of

the hidden clusters. This is a remarkable improvement comparing with existing algorithms, since these methods intend to test all possibilities following either bottom-up or top-down approach, which consequently results in an unfeasible exponential time complexity.

We also investigated the scalability w.r.t. size of dataset using a series of 20-dimensional datasets that contains 4 possibly overlapping 4-dimensional cluster. The number of points increase from 0.5 million to 2.5 millions. In all experiments our algorithm found at least 98% of the points in all hidden clusters. Figure 3 shows us our algorithm efficiently found all four clusters with remarkable accuracies. It scales linearly w.r.t. size of dataset due to the simplicity of grid-based processing.

5   Conclusions

In this paper we presented Spark2Fires, a novel parallel subspace clustering algorithm improved from Fires and implemented on top of Spark. It leverages approximation to reduce the time complexity of both base cluster generation and final cluster subspace generation, especially the latter from exponential to quadratic. Dimensionality bias is avoided since there is no global input parameter. Preliminary experimental results show that Spark2Fires has good scalability, accuracy and efficiency.

Acknowledgement. The research leading to these results has received funding from the European Union under the FP7 grant agreement no. 619633 (project ONTIC) and H2020 grant agreement no. 671625 (project CogNet)

References

1. Aggarwal, C.C., Wolf, J.L., Yu, P.S., Procopiuc, C., Park, J.S.: Fast algorithms for projected clustering. In: ACM SIGMoD Record, vol. 28, pp. 61–72. ACM (1999)
2. Huang, X., Ye, Y., Guo, H., Cai, Y., Zhang, H., Li, Y.: DSKmeans: a new kmeans-type approach to discriminative subspace clustering. Knowl.-Based Syst. **70**, 293–300 (2014)
3. Gan, G., Ng, M.K.-P.: Subspace clustering using affinity propagation. Pattern Recogn. **48**(4), 1455–1464 (2015)
4. Agrawal, R., Gehrke, J., Gunopulos, D., Raghavan, P.: Automatic subspace clustering of high dimensional data for data mining applications, vol. 27. ACM (1998)
5. Goil, S., Nagesh, H., Choudhary, A.: MAFIA: efficient and scalable subspace clustering for very large data sets. In: Proceedings of the 5th ACM SIGKDD International Conference on Knowledge Discovery and Data Mining, pp. 443–452 (1999)
6. Cheng, C.-H., Fu, A.W., Zhang, Y.: Entropy-based subspace clustering for mining numerical data. In: Proceedings of the Fifth ACM SIGKDD International Conference on Knowledge Discovery and Data Mining, pp. 84–93. ACM (1999)
7. Sequeira, K., Zaki, M.: SCHISM: a new approach for interesting subspace mining. In: Fourth IEEE International Conference on Data Mining 2004, pp. 186–193. IEEE (2004)

8. Liu, G., Li, J., Sim, K., Wong, L.: Distance based subspace clustering with flexible dimension partitioning. In: IEEE 23rd International Conference on Data Engineering 2007, pp. 1250–1254. IEEE (2007)

9. Kailing, K., Kriegel, H.-P., Kröger, P.: Density-connected subspace clustering for high-dimensional data. In: Proceedings of SDM, vol. 4. SIAM (2004)

10. Zhu, B., Mara, A., Mozo, A.: CLUS: parallel subspace clustering algorithm on spark. In: Morzy, T., Valduriez, P., Bellatreche, L. (eds.) ADBIS 2015. CCIS, vol. 539, pp. 175–185. Springer, Heidelberg (2015)

11. Assent, I., Krieger, R., Muller, E., Seidl, T.: INSCY: indexing subspace clusters with in-process-removal of redundancy. In: Eighth IEEE International Conference on Data Mining, ICDM 2008, pp. 719–724. IEEE (2008)

12. Günnemann, S., Boden, B., Seidl, T.: DB-CSC: a density-based approach for subspace clustering in graphs with feature vectors. In: Gunopulos, D., Hofmann, T., Malerba, D., Vazirgiannis, M. (eds.) ECML PKDD 2011, Part I. LNCS, vol. 6911, pp. 565–580. Springer, Heidelberg (2011)

13. Kriegel, H.-P., Kröger, P., Zimek, A.: Clustering high-dimensional data: a survey on subspace clustering, pattern-based clustering, and correlation clustering. ACM Trans. Knowl. Discov. Data (TKDD) 3(1), 1 (2009)

14. Müller, E., Günnemann, S., Assent, I., Seidl, T.: Evaluating clustering in subspace projections of high dimensional data. Proc. VLDB Endowment 2(1), 1270–1281 (2009)

15. Parsons, L., Haque, E., Liu, H., et al.: Evaluating subspace clustering algorithms. In: Workshop on Clustering High Dimensional Data and its Applications, SIAM International Conference on Data Mining, pp. 48–56. Citeseer (2004)

16. Assent, I., Krieger, R., Muller, E., Seidl, T.: DUSC: dimensionality unbiased subspace clustering. In: Seventh IEEE International Conference on Data Mining 2007, pp. 409–414. IEEE (2007)

17. Kriegel, H.-P., Kröger, P., Renz, M., Wurst, S.: A generic framework for efficient subspace clustering of high-dimensional data. In: Fifth IEEE International Conference on Data Mining, p. 8. IEEE (2005)

18. Zhu, B., Ordozgoiti, B., Mozo, A.: Psceg: an unbiased parallel subspace clustering algorithm using exact grids. ESANN (2016)

DCSA 2016 – Data Centered Smart Applications

A Smart Approach for Matching, Learning and Querying Information from the Human Resources Domain

Jorge Martinez-Gil[✉], Alejandra Lorena Paoletti, and Klaus-Dieter Schewe

Software Competence Center Hagenberg GmbH,
Softwarepark 21, 4232 Hagenberg, Austria
{jorge.martinez-gil,lorena.paoletti,kd.schewe}@scch.at
http://www.scch.at

Abstract. We face the complex problem of timely, accurate and mutually satisfactory mediation between job offers and suitable applicant profiles by means of semantic processing techniques. In fact, this problem has become a major challenge for all public and private recruitment agencies around the world as well as for employers and job seekers. It is widely agreed that smart algorithms for automatically matching, learning, and querying job offers and candidate profiles will provide a key technology of high importance and impact and will help to counter the lack of skilled labor and/or appropriate job positions for unemployed people. Additionally, such a framework can support global matching aiming at finding an optimal allocation of job seekers to available jobs, which is relevant for independent employment agencies, e.g. in order to reduce unemployment.

Keywords: e-Recruitment · Knowledge engineering · Knowledge-based technology

1 Introduction

Some of the major problems concerning the labor market are the complicated situation of the job market in many countries around the world and the increased geographical flexibility of employees. This situation makes companies to often receive a huge number of applications for every open position. Therefore, the costs of manually selecting potential candidates is usually high. For this reason, most companies would like to decrease the costs when publishing job postings

The research reported in this paper was supported by the Austrian Forschungs-forderungsgesellschaft (FFG) for the Bridge project Accurate and Efficient Profile Matching in Knowledge Bases (ACEPROM) under contract [FFG: 841284]. The research reported in this paper has been supported by the Austrian Ministry for Transport, Innovation and Technology, the Federal Ministry of Science, Research and Economy, and the Province of Upper Austria in the frame of the COMET center SCCH [FFG: 844597].

M. Ivanović et al. (Eds.): ADBIS 2016, CCIS 637, pp. 157–167, 2016.
DOI: 10.1007/978-3-319-44066-8_17

and selecting appropriate applicants from such a plethora of potential candidates [1]. It is also important to remark that unsuccessful job applicants often complain on the lack of transparency in the recruitment processes, and they often wish to receive detailed arguments, or at least, some information about the strengths and flaws of their profiles [29]. However, they do not receive any kind of feedback very often since this has to be done manually by the other part, and it is quite expensive, in terms of time and resource consumption, for the companies to do that [18].

This complicated situation leads us to think the accurate matching of curriculum vitae (CV) and job offers is very important for employers and job seekers. Therefore, the development of computational methods to optimize the recruitment processes should be of high importance in our current society [15]. Furthermore, such an approach could be beneficial for public and private employment agencies which could perform an analysis to determine the most needed qualification and training courses that would improve the skills of job seekers with respect to the market demands. As a result, a higher occupation rate could be achieved [23].

Currently, existing software solutions in this field are based on syntactic matching, i.e. for a requested profile, existing solutions check how many of the requested terms are overlapped in the candidate profile [22]. This fact ignores similarity between skills, e.g. programming skills in C++ or Java would be rated similar by a human expert [8]. Improving this primitive form of matching requires at least taking hierarchical dependencies between education or skill terms into account. To do that, various taxonomies have already been developed such as DISCO competences[1], ISCO[2] and ISCED[3]. These taxonomies play a central role in our research, since we can exploit them for achieving a more realistic mediation between open employment offers and suitable candidates. Therefore, our major contribution can be summarized as follows:

– We propose here a novel approach for the automatic matching, learning and efficient querying of information from the Human Resources (HR) domain. This approach is based on new methods that appropriately handle traditional limitations, including the uncertainty of human language, the incapability to exploit background knowledge, and the lack of a truly semantic mediation. Additionally, this approach could be of great interest for education and training institutions which could perform analysis to determine the most needed skill sets that would improve the skills of job seekers with respect to the available positions.

The rest of this paper is organized as follows: Sect. 2 describes the state-of-the-art concerning realistic matching, learning and querying information concerning HR. Section 3 describes the matching problem we are facing and why it

---

[1] http://www.disco-tools.eu.

[2] http://www.ilo.org/public/english/bureau/stat/isco/isco08/index.htm.

[3] http://www.uis.unesco.org/Education/Pages/international-standard-classification-of-education.aspx.

is relevant in this context. Section 4 explains how the HR field could be benefit from a framework for learning to rank candidates. Section 5 discusses our approach's capability for querying, and finally, we draw conclusions and put forward future lines of research.

2   State-of-the-Art

The problem of automatically matching job offers and applicant profiles has been studied in the scientific literature [2], but the complex nature of the problem we have to face, which usually involves the use of free text by employers (when writing their job offers) and by employees (when writing their application), makes developed solutions in this context unable to reach a high degree of success [18]. Some works have offered partial solutions based on the use of controlled vocabularies (i.e. ontologies) in order to fairly alleviate some problems concerning semantic mediation [5] but there are still some key challenges that should be addressed [24].

One of these most important challenges in this context is that the process of matching CVs and job offers is usually done without use of any background knowledge. Instead, overlapping information is computed. In fact, according to the researched literature, a wide range of solutions for job and profiles matching have been addressed by a variety of computational techniques, ranging from simple bipartite graph matching [7], to vector based techniques taken from classical information retrieval [6], to record matching in databases [30].

Algorithms for bipartite graph matching try to find optimal solutions when trying to maximize the number of matching relation. However, these approaches rely on assigning costs to every match between curriculum and profiles. When the costs are assigned manually, knowledge about them is completely subjective, and therefore it becomes very difficult to revise [3]. Moreover, an approach maximizing the number of matches may provide a bad service to users: for example, person P1 could have the best match for job profile J1, but she might be suggested to take job J2 just because J1 is the only available job for person P2 [13]. This means that from a strictly user-centric viewpoint, maximizing the number of matches is not the feature that could face our problem.

More sophisticated approaches are based on database techniques for record matching [12] or information retrieval [21]: feature vectors, weighted criteria, keyword-based search, assessment based on recall and precision [17]. In case of non-suitable highly ranked profiles human expertise can be used to correct inaccuracies. The problem with these techniques is that they are not appropriate for dealing with incomplete information usually present in scenarios of this kind. In fact, information about profiles is not always complete, not only because some information is unavailable, but also because some details are considered irrelevant by either the employer or the applicant. Trying to force to use an interface for entering profiles with long and tedious forms to be filled in, is the most often adopted solutions to this problem [27].

Among the problems concerning learning, the task of learning to rank has probably received the most attention in the machine learning literature in recent years. In fact, a number of different ranking problems have been introduced so far. The ranking module is one of the most important modules in a Human Resources Management (HRM) system. For a given job offer there may be hundreds or thousands of relative candidates but only a few of them are to be shown to the expert at a time. Therefore, it is very important to capture the most relevant candidates and display them to the expert. This means that the way that top candidates are presented decide the success of the HRM system, and therefore, each one of the entries is important.

The major challenge here is to use the expert behavior as a feedback. However, some researchers are skeptical about using this kind behavioral data as a feedback because there are various biases involved in taking behavior into consideration. They show that there exists some presentation bias, which is the bias involved when experts instinctively prefers some candidates in relation to others. It means some candidates are more likely to get better attention from experts and other candidates are not given the proper attention even though they are more relevant. However, it is possible to find useful strategies to solve this bias [11].

In practice, when proposing solutions concerning ranking, we think it is a good idea to consider the algorithm Okapi BM25 [25] as the baseline to compare new approaches in this field. The reason to choose an algorithm of this kind is that it is widely used by software systems to rank matching candidates according to their relevance to a given search offer. Okapi BM25 is considered the state-of-the-art among the methods using a syntactic approach [14]. Therefore, any new method in the field of automatic matching should prove its effectiveness when compared to it.

With respect to querying knowledge bases, in particular in the HR domain, the commonly investigated approach is to find the best k (with $k = 1$ in most cases) matches for a given profile (applicant profile or job offer) [4]. Though this constitutes what is commonly known as top-k-queries, a systematic investigation of such kind of queries is still missing. Top-k-queries have been thoroughly investigated in the field of databases, usually in the context of the relational data model [10], but the study of such queries in the context of knowledge bases has not yet been done. The expectation is of course, that many of the results in the relational data model can be easily adopted to this case. In particular, the focus on a single relation, i.e. the matching, as the driver for the querying, is expected to ease the extension.

In addition to top-k-queries the interest in partial orders in extended matching relations leading to skyline queries as well as global matching optimization and gap analysis place further challenges on matching-related querying of knowledge bases that have not yet been investigated. The classification of most relevant types of queries and the adaptation of corresponding state-of-the-art approaches in databases should be the emphasis in the future. The expected results are supposed to support the efficient answering of such queries.

3   Matching Information from the Human Resources Domain

In this context, semantic matching is a well know problem whereby two entities in a knowledge base are assigned a score based on the likeness of their meaning [16]. Automatically performing semantic matching is considered to be one of the pillars for many computer related fields since a wide variety of techniques rely on a good performance when determining the meaning of data they work with [19].

More formally, we can define semantic matching as a function $\mu_1 \times \mu_2 \to R$ that associates the degree of correspondence for the entities μ_1 and μ_2 to a score $s \in R$ in the range $[0, 1]$, where a score of 0 states for not correspondence at all, and 1 for total correspondence of the entities μ_1 and μ_2.

Traditionally, the way to compute the degree of correspondence between entities has been addressed from two different perspectives: using semantic similarity measures and semantic relatedness measures. Fortunately, recent works have clearly defined the scope of each of them. Firstly, semantic similarity is used when determining the taxonomic proximity between entities. For example, automobile and car are similar because the relation between both terms can be defined by means of a taxonomic relation. Secondly, the more general concept of semantic relatedness considers taxonomic and relational proximity. For example, nurse and hospital are not completely similar, but there is still possible to define a naive relation between them because both belong to the world of healthcare [19].

In most of cases, the problem to face is much more complex since it does not only involve the matching of two individual entities, but two complete documents (applicant profile or job offer). This can be achieved by computing a set of semantic correspondences between individual entities belonging to each of the two documents. A set of semantic correspondences between entities is often called an alignment. It is possible to define formally an alignment A as a set of tuples in the form $\{(id, \mu_1, \mu_2, r, s)\}$, where id is an unique identifier for the correspondence, μ_1 and μ_2 are the entities to be compared, r is the kind of relation between them, and s the score in the range $[0, 1]$ stating the degree of correspondence for the relation r.

Therefore, when matching two documents, the challenge that scientists try to address consists of finding an appropriate semantic matching function leading to a high quality alignment between these two knowledge bases. Quality here is measured by means of a function $A \times A_{ideal} \to R \times R$ that associates an alignment A and an ideal alignment A_{ideal} to two real numbers $\in [0, 1]$ stating the precision and recall of A in relation to A_{ideal}.

Precision represents the notion of accuracy, that it is to say, states the fraction of retrieved correspondences that are relevant for the matching task (0 stands for no relevant correspondences, and 1 for all correspondences are relevant). Meanwhile, *recall* represents the notion of completeness, thus, the fraction of relevant correspondences that were retrieved (0 stands for not retrieved correspondences, and 1 for all relevant correspondences were retrieved).

Applying this kind of techniques fits well in the HR scenario. The reason is that these techniques can be used for going beyond the literal lexical match of words. In this way, when analyzing the curriculum of job candidates, this kind of techniques can operate at the conceptual level when comparing specific terms (e.g., Finance) also yields matches on related terms (e.g., Economics, Economic Affairs, Financial Affairs, etc.). As another example, in the healthcare field, an expert on the treatment of cancer could also be considered as an expert on oncology, lymphoma or tumor treatment, etc. [9]. The potential of this kind of techniques is that it can support Human Resource Management when leading to a more quickly and easily cut through massive volumes of potential candidate information, but without giving up the way human experts take decisions in the real world.

4   Learning Information from the Human Resources Domain

The problem of learning can be defined as given a pair of objects (jo, ap_i) together with a measure of their suitability $y_i \in \mathbb{R}$. The goal is to learn a function $f(jo, ap_i) \approx y_i$ that approximates for every new labeled triplet example (jo, ap_i, y_i), where jo is a job offer, ap_i is a list of applicant profiles, and y_i is the associated list of scores of each ap_i for the job offer jo.

After many discussions with professionals from the Human Resources sector, we agreed this challenge has not an unique solution. The reason is that every HR professional evaluating different cases could propose different results. This makes us thinking that we should work towards an adaptive approach by means of automatic matching learning. This approach should be able to calculate the transformation cost of a given profile into a requested job offer, so that profiles with higher transformation cost should rank worse than those with lower cost. In this way, our approach should be able to replicate the results from the human experts. This means that for each person aiming to use a solution of this kind, we should train a model for capturing its know-how or preferences by means of an initial training stage. Thinking on a model of this kind is far from being trivial. However, we assume that a generic solution for this problem should be characterized by the following core attributes: (a) a base distance between sets, (b) some background knowledge to compute the replacement cost, (c) the desired cost of insertion and deletion of new elements, (d) the way to weight elements, either a multiplicative or an additive preference.

Please note that if we work with different relevant subsets (education, skills, languages, etc.) the transformations costs could be different for each subset, so the final cost should be an aggregation of the partial costs for each segmented group. Once we get a solution, the way to determine if this solution is satisfactory could be defined as the correlation between this achieved solution and an ideal one.

Concerning (a), we can formally define our distance between two sets as the minimum number of single-elements edits (i.e. insertions, deletions or substitutions) required to change one set into the other. It is very appropriate for computing the transformation costs from a CV into a job offer.

Concerning (b), setting up adequate knowledge bases that capture recruitment terminology in a precise and easily extendable way is a crucial success factor. So far, no such knowledge bases exist. However, our existing matching technology is based on valuable recruitment taxonomies. These taxonomies are structured thesaurus and vocabularies for the description of skills in different scenarios such as the education, job market and training courses respectively. These taxonomies provide us a complete skill and competence classification which is based on existing European and international standards and classifications, and therefore, represent a terminological basis for the standard description of skills, competences, occupations as well as applicant profiles, job vacancies, and job requirements, etc. or for describing professional degrees, study programs, courses, and so on. To illustrate why taxonomies are important for us, let us suppose that a job offer requests a person skilled in Java, and we have a candidate who is skilled in JavaScript. We can compute the shortest path between Java and JavaScript in the recruitment taxonomy. The transformation cost can be based on the length of this path. In this way, short paths leads to low replacement costs, and on the contrary; longer paths may lead to higher replacement costs. If there is no path between them, or even this path is not appropriate enough (i.e. too long) then we can consider insertion and deletion costs.

Concerning (c), Suitability of an applicant profile ap_i to a job offer jo needs also to consider the minimum cost of element insertions and deletions which transforms the applicant profile ap_i into the job offer jo. These costs are going to be used when an applicant profile have a different number of elements than those requested in the job offer or computing the replacement cost between elements is not possible. The computation of these costs is of vital importance because it helps us to characterize the behavior of the people who was involved in the training stage. Insertion cost is an estimation of how much it could cost to a potential candidate to acquire an element requested by the job offer.

Deletion cost is an estimation about the impact of having a not requested element. For example, an expert could think that candidates holding not requested elements could be unhappy, unmotivated, could request a higher salary or be willing to leave the company in a short period of time. The penalty to be applied can be high, if the person in the training phase tends to penalize over-qualification, null if the person does not care about additional (although not requested) elements, or even negative, if the person training the model thinks that additional elements are far from hurting. It is also important to note that we cannot have an unique value for insertions and deletions costs. For instance, it is much more expensive (in terms of effort, time and money) acquiring a new university degree that some certain level of mastery in a programming language or technology.

Concerning (d) the weighting schema is the way a person could increase or decrease the importance of the elements within a given set. Considering a weighting schema is important because it allows job recruiters giving more importance to some facts like years of experience, level of mastery or simply stating priorities for filling a position.

4.1 Querying Information from the Human Resources Domain

One of the main requirements from the HR application domain leads to queries on a knowledge base of job offers and candidate CVs. Ignoring the inherent inferential capability given by knowledge bases. Each knowledge base is also a database in the sense that there is a schema, i.e. the concepts and roles in the TBox, and a set of instances, i.e. the ABox. Therefore, adopting database technology as key method to address the querying problems is a natural idea [23].

In database technology effective and efficient query processing is a core area with a tradition since decades. Recently, two classes of queries, top-k-queries and skyline queries have attracted the interest of researchers [26]. For top-k-queries assume that a query q produces an answer set $\mathcal{A}$ that is totally ordered. Then a query top-$k(q)$ will select the k largest elements of $\mathcal{A}$ as the answer. While performing a sorting operation and a cut-off of the largest k elements are straightforward in theory, the key problem with top-k-queries is efficiency on very large databases, for which supporting data structures and rewriting techniques that enable the computation of the k largest answers without computing first all answers. Similarly, skyline queries ask for all maximal elements in an answer set $\mathcal{A}$ to a query q, where $\mathcal{A}$ is assumed to be partially ordered.

We think that top-k- and skyline queries are essential for the core of matching related queries, where the (partial) order is defined by the matching measures. In case of simultaneous use of several matching measures a partial order may result. Therefore, the key research question is to adopt the solutions from database technology to the area of knowledge bases, which boils down to investigating efficient storage of the ABox including matching measures. For the data structures supporting the subsumption hierarchy it is envisioned that rings and spiders [20] known from network databases and revived in object-oriented databases can be adopted. These structures are known for excellent performance in support of queries that exploit hierarchical data structuring. Furthermore, indices based on partial fractions may also be exploited for this purpose [28].

It is further anticipated that skyline queries will also play an important role for gap analysis, which should result in minimally enlarged filters that guarantee improved matching results. That is, we have to exploit a partial order on filters for such queries. The enlargement itself requires for data structures supporting neighborhoods, which will be a new notion that has to be defined and for which suitable storage representations have to be found. With such extensions it should be possible to exploit state-of-the-art techniques for skyline queries to support the application needs. Furthermore, specific query optimization techniques will be needed.

The adaptation and extension of query optimization is also the method that is needed to support global matching with respect to some optimization criteria. As the optimization criteria will lead again to a partial order, this gives another class of skyline queries, so the remaining problem is efficiency, for which query optimization is due.

5   Discussion

We think our approach lead to a number of qualitative advantages over the state-of-the-art in this field. These advantages are in the direction of those mentioned in [18]. In fact, we can summarized them in the following four major points:

1. Our approach for realistic matching learning can help players from the HR industry to go beyond syntactical matching of job offers and applicant profiles. This represents a great advantage over the current state-of-the-art since our approach tries to give more opportunities to the good job candidates, but also allows job recruiters to identify potential talent which otherwise may remain blurred among such a plethora of applicants profiles.
2. Our approach can help to eliminate the need for job recruiters to have deep and specialized knowledge within an industry. This is mainly due to this approach is able to model knowledge from a lot of industrial domains. Then this knowledge can be used as a support when performing matching process so that the results can be very similar to those produced by an expert from that field.
3. Our approach can provide feedback to the applicants that did not get the job. The matching process is traceable and this means that some interesting reports can be automatically delivered to the applicants. These reports can help these applicants to determine the reasons they were not selected for the job position as well as to assess their strengths and weaknesses when applying for similar jobs in the future.
4. Our approach allows to leveling the odds for those job applicants with less ability when preparing their resumes. The reason is that an algorithm will perform the matching process automatically. The result from this process is independent of the way the curriculum is presented. Therefore, this technique helps to promote equal opportunities.

6   Conclusions

We have presented a novel approach for the timely, accurate and mutually satisfactory mediation between open employment offers and suitable candidates. The rationale behind this research approach is to facilitate public and private recruitment agencies as well as employers and job seekers around the world to reduce the costs and time to find relevant matches between job offers and applicant profiles.

The major conclusion we can extract is that an approach of such kind may be able overcome the traditional limitations in this field. Concerning the uncertainty when dealing with natural language: we propose to describe either job offers and applicant profiles using a common vocabulary. This fact avoid the problem of dealing with different representations of a same concept. Concerning the incapacity of current approaches to exploit background knowledge, we propose to get benefit from external knowledge bases that support our work within the process of making estimations about the cost of acquiring a new skill or competence. Concerning achieving a truly realistic mediation, we propose to go beyond syntactic matching of job offers and applicant profiles. Additionally, some practical impact can be envisioned, e.g. capability to reduce unemployment.

As a future work, we plan to develop a benchmark data set in order to assess the quality of the new matching proposals based on this and other different paradigms. The importance of this benchmark data set is high since researchers need some data to test their solutions. One solution could consists of automatically generated random data sets. The problem is that randomly generated data sets are not useful since they do not have realistic distributions, and it is not always easy to understand whether the achieved results are meaningful and correct. Real data is much better in this particular application scenario. For these reasons, we are working towards a benchmark data set that reflects the way real HR experts take decisions in real scenarios and allows research community testing and training new approaches under truly realistic conditions.

References

1. Bizer, C., Heese, R., Mochol, M., Oldakowski, R., Tolksdorf, R., Eckstein, R.: The impact of semantic web technologies on job recruitment processes. Wirtschaftsinformatik 2005, pp. 1367–1382 (2005)
2. Bradley, K., Smyth, B.: Personalized information ordering: a case study in online recruitment. Knowl. Based Syst. **16**(5–6), 269–275 (2003)
3. Cali, A., Calvanese, D., Colucci, S., Di Noia, T., Donini, F.M.: A logic-based approach for matching user profiles. In: Negoita, M.G., Howlett, R.J., Jain, L.C. (eds.) KES 2004. LNCS (LNAI), vol. 3215, pp. 187–195. Springer, Heidelberg (2004)
4. Chakrabarti, K., Ortega-Binderberger, M., Mehrotra, S., Porkaew, K.: Evaluating refined queries in top-k retrieval systems. IEEE Trans. Knowl. Data Eng. **16**(2), 256–270 (2004)
5. Colucci, S., Di Noia, T., Di Sciascio, E., Donini, F.M., Mongiello, M., Mottola, M.: A formal approach to ontology-based semantic match of skills descriptions. J. UCS **9**(12), 1437–1454 (2003)
6. Faliagka, E., Tsakalidis, A.K., Tzimas, G.: An integrated E-recruitment system for automated personality mining and applicant ranking. Internet Res. **22**(5), 551–568 (2012)
7. Farber, F., Weitzel, T., Keim, T.: An automated recommendation approach toselection in personnel recruitment. In: AMCIS 2003, p. 302 (2003)
8. Garcia Sanchez, F., Martinez-Bejar, R., Contreras, L., Fernandez-Breis, J.T., Castellanos Nieves, D.: An ontology-based intelligent system for recruitment. Expert Syst. Appl. **31**(2), 248–263 (2006)
9. Chaves-Gonzalez, J.M., Martinez-Gil, J.: Evolutionary algorithm based on different semantic similarity functions for synonym recognition in the biomedical domain. Knowl. Based Syst. **37**, 62–69 (2013)
10. Ilyas, I.F., Aref, W.G., Elmagarmid, A.K.: Supporting top-k join queries in relational databases. VLDB J. **13**(3), 207–221 (2004)
11. Joachims, T.: The value of user feedback. In: Clough, P., Foley, C., Gurrin, C., Jones, G.J.F., Kraaij, W., Lee, H., Mudoch, V. (eds.) ECIR 2011. LNCS, vol. 6611, p. 6. Springer, Heidelberg (2011)

12. Kessler, R., Bechet, N., Roche, M., Torres-Moreno, J.M., El-Beze, M.: A hybrid approach to managing job offers and candidates. Inf. Process. Manage. **48**(6), 1124–1135 (2012)

13. Kuokka, D., Harada, L.: Integrating information via matchmaking. J. Intell. Inf. Syst. **6**(2/3), 261–279 (1996)

14. Lv, Y., Zhai, C.X.: A log-logistic model-based interpretation of TF normalization of BM25. In: Baeza-Yates, R., de Vries, A.P., Zaragoza, H., Cambazoglu, B.B., Murdock, V., Lempel, R., Silvestri, F. (eds.) ECIR 2012. LNCS, vol. 7224, pp. 244–255. Springer, Heidelberg (2012)

15. Malinowski, J., Keim, T., Wendt, O., Weitzel, T.: Matching people, jobs: a bilateral recommendation approach. In: HICSS 2006 (2006)

16. Martinez-Gil, J., Aldana-Montes, J.F.: Reverse ontology matching. SIGMOD Rec. **39**(4), 5–11 (2010)

17. Martinez-Gil, J., Aldana-Montes, J.F.: Evaluation of two heuristic approaches to solve the ontology meta-matching problem. Knowl. Inf. Syst. **26**(2), 225–247 (2011)

18. Martinez-Gil, J.: An overview of knowledge management techniques for e-recruitment. JIKM **13**(2), 1450014 (2014)

19. Martinez-Gil, J.: Automated knowledge base management: a survey. Comput. Sci. Rev. **18**, 1–9 (2015)

20. Mylopoulos, J., Brodie, M.L.: Knowledge bases and databases: current trends and future directions. In: Karagiannis, D. (ed.) IS/KI 1990 and KI-WS 1990. LNCS, vol. 474, pp. 153–180. Springer, Heidelberg (1991)

21. Meo, P., Quattrone, G., Terracina, G., Ursino, D.: An XML-based multiagent system for supporting online recruitment services. IEEE Trans. Syst. Man Cybern. Part A **37**(4), 464–480 (2007)

22. Mochol, M., Wache, H., Nixon, L.J.B.: Improving the accuracy of job search with semantic techniques. In: Abramowicz, W. (ed.) BIS 2007. LNCS, vol. 4439, pp. 301–313. Springer, Heidelberg (2007)

23. Paoletti, A.L., Martinez-Gil, J., Schewe, K.-D.: Extending knowledge-based profile matching in the human resources domain. In: Chen, Q., Hameurlain, A., Toumani, F., Wagner, R., Decker, H. (eds.) DEXA 2015. LNCS, vol. 9262, pp. 21–35. Springer, Heidelberg (2015)

24. Rácz, G., Sali, A., Schewe, K.-D.: Semantic matching strategies for job recruitment: a comparison of new and known approaches. In: Gyssens, M., et al. (eds.) FoIKS 2016. LNCS, vol. 9616, pp. 149–168. Springer, Heidelberg (2016). doi:10.1007/978-3-319-30024-5_9

25. Robertson, S.E., Walker, S., Hancock-Beaulieu, M.: Experimentation as a way of life: okapi at TREC. Inf. Process. Manage. **36**(1), 95–108 (2000)

26. Soliman, M.A., Ilyas, I.F., Ben-David, S.: Supporting ranking queries on uncertain and incomplete data. VLDB J. **19**(4), 477–501 (2010)

27. Straccia, U., Tinelli, E., Colucci, S., Di Noia, T., Di Sciascio, E.: A system for retrieving top-k candidates to job positions. In: Description Logics 2009 (2009)

28. Theobald, M., Weikum, G., Schenkel, R.: Top-k query evaluation with probabilistic guarantees. In: VLDB, pp. 648–659 (2011)

29. Thielsch, M.T., Traumer, L., Pytlik, L.: E-recruiting and fairness: the applicant's point of view. Inf. Technol. Manage. (ITM) **13**(2), 59–67 (2012)

30. Tinelli, E., Cascone, A., Ruta, M., Di Noia, T., Di Sciascio, E., Donini, F.: An innovative semantic-based skill management system exploiting standard SQL. In: ICEIS (2), pp. 224–229 (2009)

Smart Modeling for Lightweight
Mobile Application Development Methods

Eiman Alsabi and Ajantha Dahanayake(✉)

College for Women, Prince Sultan University, King Abdullah Road,
Riyadh, 11585, Saudi Arabia
Alsabi.eman@gmail.com, AjanthaDahanayake@yahoo.com

Abstract. In recent years, the rapid proliferation of mobile devices has led to a sharp rise in the number of available mobile applications. In turn, this has led to the need for more structured mobile application development methods, mainly due to the peculiarities of the mobile development process such as the rapid change of the stakeholder's requirements and the need for early feedback from the users. In addition, mobile applications are characterized by redundant source code, which deteriorate the performance of the mobile device, due to the fact that developers do not follow a clear structured approach. To this end the scope of the paper is two-fold. First, it briefly reviews the existing methods and attempts to show their limitations. Thereafter, the lightweight Mobile Application Development Method (MADeM) is proposed. MADeM is built upon the principles of the methodology engineering framework and utilizes the SMART models. The aim of MADeM is to provide a systematic and structured development approach, which can lead to more effective mobile applications development by removing the accumulation of the redundant software code.

Keywords: Mobile application development · Method engineering · Agile methods · Smart modeling

1 Introduction

In recent years, mobile computing has undergone many advances, leading to the rapid increase in the number of mobile devices available in the market, such as mobile phones and tablets [1]. As a result, mobile applications, which run on those devices, have become pervasive in the daily life and have the potential to assist in each aspect of the human life such as the social, leisure, health and work [2]. Their popularity becomes evident from the number of applications available on the App stores. Indicatively, it is reported that as of July 2015 the two major App stores, Google Play and Apple App Store offer more that 3 M applications[1].

The mobile applications are surrounded by a fast changing environment where the customer requirements and expectations are anticipated to change frequently [1]. Moreover, there is a certain number of constraints which shape the development of mobile applications, such as the limited resources of the mobile devices, the wireless

---

[1] http://www.statista.com/statistics/276623/number-of-apps-available-in-leading-app-stores/.

© Springer International Publishing Switzerland 2016
M. Ivanović et al. (Eds.): ADBIS 2016, CCIS 637, pp. 168–179, 2016.
DOI: 10.1007/978-3-319-44066-8_18

communication problems, the heterogeneity of platforms and the short time-to market requirements [3]. In addition, the rapidly evolving market requires lightweight development processes which are able to respond quickly to emerging technologies and new requirements [4]. To further challenge the current state of affair, Gartner a leading research analysis company, reports that there is a rapid increase in the demand for specialized mobile application development methods, since the traditional software development practices fails to meet the requirement of the mobile applications [5].

As a response, several mobile application development methods have emerged, such as the Mobile-D [6], the RaPiD7 [7], the Hybrid Methodology Design [8] and the Mobile Application Software Agile Methodology (MASAM) [9]. Nevertheless, these methods lack to present a systematic and structured way for supporting the development process. At the same time they may rely on extensive documentation, which is often time consuming and distracts the development team from the actual development tasks. As a result, there is no available development method, which is well established and widely used by the developers. In turn this leads to the fact that mobile applications often contain more code than required leading to overuse of the mobile device resources.

To this end the paper proposes the constitution of the lightweight Mobile Application Development Method (MADeM). The foundation of MADeM lies in the methodology engineering framework, which puts forward the notion that every method should define clearly the way of thinking, the way of modelling, the way of working, the way of controlling and the way of supporting [10, 11]. Thereby, the method attempts to provide a structured and systematic development process, which in turn can facilitate the reduction of the redundant source code. Additionally, MADeM leverages the SMART models, namely a collection of specific models, which are used by the development team in specific phases throughout the development process [12, 13]. These are: The Informative Model, the Use Case Diagram, the Component Diagram, the Class Diagram, the Activity Diagram, the Sequence Diagram, the Model-View-Controller Pattern and the User View Model.

The remainder of the paper is structured as follows: Sect. 2 provides a brief review of the most prominent mobile application development methods. Next, the methodology engineering framework is explained, based on which MADeM is built. Section 4, discusses the SMART models and defines the components of the proposed MADeM method. Thereafter, Sect. 4 reports on a case study demonstrating how MADeM can be applied in practice and further compares and analyses the proposed method against the existing mobile application development methods. Finally, Sect. 5 concludes the paper and provides concrete future research directions.

2   Mobile Application Development Methods

According to [1, 4, 14], the most prominent mobile application development approaches reported in the scientific literature are: Mobile-D [6], Rapid Production of Documentation–7 steps (RaPiD7) [7], Hybrid Methodology Design [8], and the Mobile Application Software Agile Methodology (MASAM) [9].

Mobile-D: Mobile-D is based on the Agile approach, proposed by Abrahamsson et al. [6]. It lays its foundations on the Extreme Programming (XP) and the Rational Unified

Process (RUP) development methods. Mobile-D splits the project development into five phases: Explore, Initialize, Productionize, Stabilize, System Test and Fix, and each one of them is associated to specific stages and activities [6]. According to the founders of Mobile-D, the method can identify and solve technical problems at an early stage, and further promote the shared responsibility, the efficient information sharing and a constant development rhythm. Nevertheless, Mobile-D relies on the plain use of Unified Modelling Language (UML) models and extensive documentation, which may be time consuming for the development team [15, 16].

RaPiD-7: RaPiD7 is developed by Nokia during 2000–2001 as a response to the need for creating and sharing understanding and subsequently documenting the created understanding [7]. RaPiD7 introduces a three-layer structure. The first one is the project layer, which dictates the way the human interaction and the joint decision-making takes place in order to identify the cases for applying RaPiD7. Then the case layer, determines the way the cases, such as the documents, will be created in successive workshops. Finally, the workshop layer defines how the actual work will be carried out [7]. The anticipated benefit of using RaPiD7 is that the documentation, which is produced, is small in size and thus is easier to maintain.

Hybrid Methodology Design: Hybrid Methodology Design, of Rahimian and Ramsin [8], is based on the Adaptive Software Development (ASD) [17] and the New Product Development (NPD) [18] and proposes a top-down iterative incremental process for mobile applications development. The project development process begins with the generation of the idea. Then, the project initiation phase starts including a preliminary business analysis. Next comes the Design phase during which, the high-level architecture is designed followed by the creation of the detailed design. The third phase involves the actual development including the quality assurance and the market testing. Finally, commercialization activities take place. However, the method is only defined at a high-level and no specific tasks are described for each of the phases [14]. Moreover, the effectiveness of the Hybrid Methodology Design has not been validated with a use of any case study.

MASAM: The Mobile Application Software Agile Methodology (MASAM) [9] is influenced by XP and RUP methods and shares characteristics with Mobile-D. However, it pays more attention to project management [14]. The method defines four phases [9]: Preparation phase, where the team communicates with the client to obtain a high-level idea of the product to be developed; Embodiment phase, during which the initial requirements are captured and the client is presented with a high-level prototype or a mock-up of the final product; Product Development phase, where the actual implementation takes place; Commercialization phase, which involves customizing the product for specific countries and releasing beta versions for capturing the user's experience. Similar to Hybrid Methodology Design, there are no details about the concrete tasks involved in each phases and additionally there are no case studies reported in the literature that demonstrate the feasibility of MASAM.

As seed in Table 1, Mobile-D is reportedly applied in 16 case studies, while MASAM and Hybrid Methodology Design have no reported case studies [7]. Thus, Mobile-D appears to be the most appropriate choice for the development of mobile applications and as such the proposed MADeM is partially based on Mobile-D. However, Mobile-D is influenced by a combination of different methods (XP, RUP), and as a result, it relies on the modeling approaches of UML [15]. However, as also indicated by Inukollu et al. [19], the plain UML models are not adequate to address the needs of mobile application development. Additionally, the extensive documentation, that Mobile-D relies on, may be time-consuming and lead to delays in the development process [16]. By contrast, the proposed method attempts to overcome this shortcoming by applying the SMART models. The proposed method is structured based on the methodology engineering framework [10, 11].

Table 1. Mobile application development methods

Method	Empirical evidences	Limitations
Mobile-D	16	Extensive Documentation
		Sole use of UML modelling
RaPiD-7	1	Main focus on the documentation
Hybrid Methodology Design	0	No detailed instructions of the concrete phases of the method
MASAM	0	No detailed instructions of the concrete phases of the method

3 A Novel Mobile Application Development Method Using the SMART Modelling Approach

The Mobile Application Development Method (MADeM), proposed in this paper, is not developed from the ground up. As highlighted by Hong et al., the "new methods" are largely influenced by other methods and techniques [20]. Therefore, the method seeks to exploit the agile development approach [21], since agile approaches can efficiently and rapidly respond to changes by adopting short development cycles and anticipating modifications in the requirements [21]. In particular, the proposed method is influenced by Mobile-D. Nevertheless, as discussed in Sect. 2, MADeM rather than adopting solely the UML models, it proposes the use of SMART models as a mean to sufficiently address the needs of mobile application development regarding the clear definition of the requirements and the provision of meaningful information about the application. In addition, with the use of the SMART models, it attempts to decrease the redundant documentation and thereby minimize the overhead for the development team. The MADeM method is constructed on the basis of the methodology engineering framework.

3.1 The Methodology Engineering Framework

The methodology engineering framework proposed by Wijers [11] and Dahanayake [10] has been widely used in PhD researchers who developed methods in the systems development field[2]. The framework consists of the following five "ways":

---

[2] http://repository.tudelft.nl/.

The **way of thinking** visualizes the essential of systems development methods regarding the information system functionality and role in the environment. It verbalizes assumptions and viewpoints of the method on the kinds of problem to be solved, solutions, and models, in other words it describes the philosophy.

The **way of working** structures the way in which an information system is developed. It defines the possible tasks, including sub-tasks, and orderings of tasks, to be performed as part of the development process. It furthermore provides guidelines and suggestions on how these tasks should be performed.

The **way of modelling** provides an abstract description of the underlying modelling concepts together with their interrelationships and properties. It structures the models, which can be used in the information system development, depending on the type of models required for problem specification and solution finding.

The **way of controlling** deals with managerial aspects of information system development. It includes set of directives and guidelines, quality and progress control, and evaluation of plans.

The **way of supporting** of a method, refers to the support of the method by tools. A generally accepted name for computer based tools for methods are: CASE tools.

Next sections describe how MADeM is constructed based on the above-mentioned framework (Fig. 1).

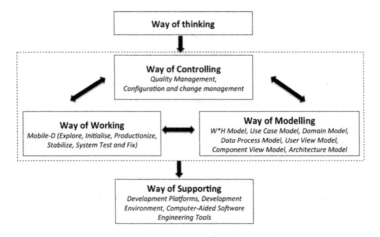

Fig. 1. MADeM and the methodology engineering framework

3.2 The Way of Thinking

MADeM's way of thinking primarily focuses on the development of mobile applications. It provides guidelines, practices and principles to make the approach lightweight and easy to use.

3.3 The Way of Modelling

As the mobile application development needs to be flexible, the use of all the UML modeling concepts and techniques is not recommended [19]. Therefore, a smart modeling technique is introduced, which is based on the SMART models defined in [12, 13].

3.3.1 SMART Models

The SMART models stand for simple, meaningful, adequate, realistic and track-able model [12, 13]. The characteristics of the *SMART* models are:

Simple: The simplicity of the models can be derived from explicit consideration for purpose. The models explicitly answer the questions of how, when, what, where and by whom the purpose will be achieved, limiting the ambiguity.

Meaningful: The models provide meaningful information about the target to be accomplished by evaluating the progress that has been achieved.

Adequate: This characteristic ensures that the models are simple, yet they are well-formed and purposeful.

Realistic: It answers questions regarding the feasibility of the models, such as whether the models are doable, worthwhile, can be delivered on time, match the needs and efforts of the community of practice and whether they are capable of evolving.

Trackable: The models provide limits, deadlines and benchmarks. They state when, how long and when to terminate. They provide the sequential order of activities to perform.

By following the conceptual modeling theory presented in [12, 13], the following models are identified as SMART models.

1. **The Informative Model**. This model provides the required information regarding the mobile application development project. Specifically, it informs the users, guides and steers them by adjusting the information stages and levels (Table 2). In the case of the mobile applications modeling, the informative model must state in an understandable way what the application is, and additionally must describe the inputs and the outputs of the application. The informative model for mobile application development is given in Table 2 and has been constructed using the W*H framework specifications provided in [12, 13]. This framework provides the informative model of the application and explains the application in terms of the 23 questions defined in [12, 13].
2. **The Use Case Diagram.** This diagram, is selected for user requirement modeling. Use cases illustrate the required functionality of an application and each use case represents the interaction between the user and the application or the system.
3. **The Component-View Diagram**. This diagram is used to depict the major components of the application with their relations. The componentization of the application is promoted in order to enhance the maintainability and the reusability of the software code of the application.

4. **The Class Diagram**. This diagram represents the domain model and contains the actual data model of the software application, the classes and the relation between them.

Table 2. Informative model for mobile applications [13]

Application	Application	Name		
		Wherefore?		
Concept	Ends		Why?	
		Purpose	Where to?	
			For When?	
			For which reason?	
		Wherewith?		
Content	Supporting means	Application Domain	Application are	Wherein?
			Application case	Wherefrom?
			Problem	For what?
			Organizational Unit	Where
			Triggering Event	Whence
			IT	What
				How
Annotation	Source	Where of?		
		Party	Supplier	By whom?
			Consumer	To whom?
			Producer	Whichever?
		Activity	Input	What in?
			Output	What out?
Added Value	Surplus Value	Worthiness?		
		Context	System Context	What at?
			Story Context	Where about?
			Coexistence context	Wither?
			Time Context	When?

5. **The Activity Diagram**. This diagram is used for interaction and process modelling. In particular, the Activity Diagram represents the workflow of actions and activities and provides support for concurrency, iteration, and choice.
6. **The Sequence Diagram**. This diagram is used to depict the detailed interactions between the objects of the software system in sequential order. Each sequence diagram is used to represent a single scenario that is part of the use-case model.
7. **The Model-View-Controller (MVC).** Pattern for software architectural modeling. The MVC divides the software application into its interconnected parts. These parts are:

- Model: It stores data and serves the view based on the incoming command from the controller.
- Controller: It passes the commands from the view to the model and displays the data to the view that have been processed by the model.
- View: It takes the user inputs and displays the output.

8. **The User View Model.** This model represents how the user interacts with the software application, what are the inputs and how the user responds to the output of the application. The User View modeling can assist in capturing the user requirements, avoiding premature commitment to specific layouts and widgets, and making explicit the relationships between the different parts of an interface and their roles.

In total the 8 SMART models are: Informative Model, Use Cases, Component View Model, Class Diagram, Activity Diagram, Sequence Diagram, Model-View-Controller Pattern and the User View Model.

3.4   The Way of Controlling

The way of controlling involves the management facilities provided by the method in order to handle the time, the cost and the quality aspect of the software development process. MADeM adopts the way of controlling of the Mobile-D method.

At the beginning of the project, a concrete project plan is constructed to include the project phases, with the involved tasks in each phase, together with the deadline to complete each task. In the development process, the plan will be updated to reflect each time the requirements and the time frame of the project.

The use of a product log ensures proper configuration and change management. The product log contains all the requirements of the project as a list of features. Each feature is divided into tasks, called task cards. After each release cycle, the product log is updated to reflect the current features with the task cards will be implemented. In parallel, the daily software development updates in the projects code is tracked using version control system (VCS) and each change is called a version. VCS helps in code tracking, change management and version control.

The MADeM method supports the Test-Driven Development (TDD) process. TDD is a software development practice, in which unit test cases are incrementally written prior to code implementation [22]. Unit testing is performed to detect faults that pertain to the individual units, independent of the rest of the system. The overall quality of the delivered software is also ensured by the stakeholder's approval after each release cycle. The short release cycles guarantees the regular involvement of the clients in the quality process.

3.5   The Way of Working

The way of working refers to the way the development process is carried out including the involved phases and tasks. MADeM adopts the way of working of Mobile-D method, which consists of the following five phases:

1. The Explore phase, prepares the foundation of the development process. It comprises three stages: stakeholder establishment, project establishment and scope definition. The Informative Model is used during this phase in order to establish a common understanding of the application development project, among the stakeholders of the project. Additionally, the User View Model offers an insight into the final user interface.

2. The Initialize phase, consists of three sub-phases: trial day, initial planning and project setup. The developers meet regularly the customers, in order to understand the requirements and decide on the technology to be used during the development. The User View Model is updated to reflect the status of the development process.

3. The Productionize phase, includes the actual implementation. It is divided into Planning day, Release day and Working days.

   - The Planning days aim at analyzing enhancing and prioritizing the requirements, planning the contents of current iteration and preparing the acceptance test cases that will be used on Release day.
   - The Working days implement functionalities in Test-Driven Development (TDD) fashion.
   - Finally, in Release days, a working version is released for acceptance testing from client using the test cases developed during the Planning days.
     Besides the Informative and the User View Model, during the Productionize phase, the development team uses the Use Case, the Component, the Class the Activity and the Sequence Diagrams as well as the MVC Pattern in order to carry out the implementation and align the understanding of the team members.

4. The Stabilize phase involves the integration of smaller subsystems, developed by different teams, into a single project. In addition, the final documentation is produced and delivered to the project stakeholders.

5. The System Test and Fix phase involves thorough testing of the whole system in order to detect and fix any defects in the produced software.

3.6 The Way of Supporting

The way of supporting involves the software tools, which are available to the developers for the creation of the mobile application. MADeM does not provide any tools but encourages the use of established and widely used technologies.

Specifically, MADeM supports all major mobile operating systems such as: Android, iOS and Windows Phone. Regarding the Software Development Kits (SDKs), the programmers can use the iOS SDK with XCode, the ADTP plugin for Eclipse programming environment or the .NET compact framework with Microsoft. The GIT and the Subversion (SVN) are recommended for version control management.

4   Analysis and Discussion

4.1   Applying MADeM

In order to demonstrate the feasibility of the proposed MADeM method, a case study has been conducted by implementing the "DiabeticMedicare", diabetes management mobile application. The application assists the diabetic patients in the efficient management of the diabetes. Specifically, the application is able to record the levels of glucose in the blood, report the values to the doctor, remind the patients about the medication and provide the patients with useful information and advice about diabetes.

During the development process, the way of working of MADeM was applied, namely the Explore, the Initialize, the Productionize, the Stabilize and the System Test and Fix phase. Regarding the way of modelling, the SMART models have been used in order to capture the requirements and subsequently to create the architecture and the detailed design of the application. To ensure the control of the time schedule of the project, a product log was maintained where all the implementation tasks were tracked. Additionally, version control system was used to keep track of the various versions of the software and TDD was applied to ensure the quality of the produced code. With respect to the way of supporting, the Android SDK along with the Eclipse IDE were used.

In order to validate the "DiabeticMedicare" and the proposed MADeM a user survey has been carried out, consisting of 14 questions. It evaluates the usability of the application by measuring the efficiency, the effectiveness and the degree of user satisfaction [23]. The survey has been performed in the form of questionnaires, which have been distributed to 30 diabetic patients, out of which 27 responded, aging from 15 to 80 years old. The results are presented in Table 3 (The scale is from 5-1, 5–most positive 1–most negative response).

Table 3.   Results of the usability survey

Usability Metrics	5	4	3	2	1
Efficiency	62%	27.8%	9.3%	0%	0.9%
Effectiveness	68%	14.2%	3%	3.7%	11.1%
User Satisfaction	77.8%	13.9%	4.6%	0%	3.7%

4.2   Comparative Analysis of MADeM

Several mobile application development methods were reviewed in Sect. 2. These methods follow various processes during the development of applications and may have different focus. For instance, MASAM is influenced by the Agile processes, RUP and XP, whereas RaPiD7 primarily focuses on optimizing the produced documentation [14]. Based on the review carried out in Sect. 2 and the available case studies for each method, Mobile-D proves to be the most popular method. Therefore, MADeM is primarily influenced by Mobile-D. However, Mobile-D relies on extensive documentation, which can be time consuming and lead to delays in the development process [16].

Regarding the way of modelling, most of the previous methods promote the use of UML models. However, the sole use of UML models do not appear to be optimized for mobile application development with respect to establishing a clear understanding about the context of the application. Besides that, their excessive use can be time consuming. By contrast the proposed lightweight Mobile Application Development Method (MADeM) propose the use of 8 SMART models. These models are used in specific phases throughout the development of the application, as reported in Sect. 3.3. Subsequently, MADeM has the potential to shorten the time to market of the final product, a crucial factor for the mobile applications. Furthermore, the clear understanding of the stakeholder's requirements from the very beginning, can lead to the reduction of the unnecessary software code in the application, which in turn can increase the performance of the mobile application.

5 Conclusion and Future Work

This paper presented the novel lightweight Mobile Application Development Method (MADeM). MADeM addresses the limitations of the existing mobile application development methods: The methods are based on the use of generic scope of UML modelling and extensive documentation, resulting in the underutilization of the development process. MADeM overcomes the existing limitation with the use of the SMART models,. The case of the "DiabeticMedicare" application briefly described how MADeM can be applied in practice. In order to further strengthen the applicability of the method, the perspective of the developers may also be captured. Thus, future work involves carrying out an empirical study in which software developers are asked to use the MADeM method and report their development experience. Additionally, a toolkit can be constructed to enable the automatic generation of source code out of the SMART models and thus further speed up the development process.

References

1. Harleen, X.W.F.: Adopting an agile approach for the development of mobile applications. Int. J. Comput. Appl. **94**(17), 43–50 (2014)
2. Wac, K., Ickin, S., Hong, J., Janowski, L., Fiedler, M., Dey, A.: Studying the experience of mobile applications used in different contexts of daily life. In: 1st ACM SIGCOMM Workshop on Measurements Up the Stack, pp. 7–12. ACM, New York (2011)
3. Wasserman, A.I.: Software engineering issues for mobile application development. In: FSE/SDP Workshop on Future of Software Engineering Research, pp. 397–400. ACM, Santa Fe (2010)
4. Corral, L., Sillitti, A., Succi, G.: Software development processes for mobile systems: Is agile really taking over the business? In: 1st International Workshop on the Engineering of Mobile-Enabled Systems, pp. 19–24. ACM, San Francisco (2013)
5. Lee, C: Gartner Says Traditional Development Practices Will Fail for Mobile Apps. Technical Report, Gartner (2014)

6. Abrahamsson, P., Hanhineva, A., Hulkko, H., Ihme, T., Jäälinoja, J., Korkala, M., Koskela, J., Kyllönen, P., Salo, O.: Mobile-D: an agile approach for mobile application development. In: 19th Annual ACM SIGPLAN Conference on Object-oriented Programming Systems, Languages, and Applications, pp. 174–175. ACM, Vancuver (2004)

7. Dooms, K., Kylmäkoski, R.: Comprehensive documentation made agile–experiments with RaPiD7 in Philips. In: Bomarius, F., Komi-Sirviö, S. (eds.) Product Focused Software Process Improvement, pp. 224–233. Springer, Heidelberg (2005)

8. Rahimian, V. Ramsin, R.: Designing an agile methodology for mobile software development: A hybrid method engineering approach. In: 2nd International Conference on Research Challenges in Information Science, pp. 337–342. IEEE Press, Marrakech (2008)

9. Jeong, Y.J., Lee, J.H., Shin, G.-S.: Development process of mobile application SW based on agile methodology. In: 10th International Conference on Advanced Communication Technology, pp. 362–366. IEEE press, Gangwon-do (2008)

10. Stojanovic, Z., Dahanayake, A., Sol, H.: A methodology framework for component-based system development support. In: 6th International Workshop on Evaluation of Modeling Methods in Systems Analysis and Design, pp. 1–12. IGI Global, Interlaken (2000)

11. Wijers, G.M.: Modelling support in information systems development. Delft University of Technology (1991)

12. Dahanayake, A., Thalheim, B.: Informative models of smart and mobile services In: Transactions on Large-Scale Data and Knowledge-Centered Systems (2016, to be published). Special Issue on Advanced Techniques for Cloud Data Management

13. Dahanayake, A., Thalheim, B.: W * H: the conceptual model for services. In: Thalheim, B., Schewe, K.-D., Prinz, A., Buchberger, B. (eds.) Correct Software in Web Applications and Web Services, pp. 145–176. Springer, Switzerland (2015)

14. Flora, H.K., Chande, S.V.: A review and analysis on mobile application development processes using agile methodologies. Int. J. Res. Comput. Sci 3(4), 9 (2013)

15. Ruparelia, N.B.: Software Development Lifecycle Models. SIGSOFT Softw. Eng. Notes 35(3), 8–13 (2010)

16. Zlatko, S.: A proposal of an ontology-based methodological framework for multi-platform mobile applications development. University of Zagreb (2014)

17. Awad, M.A.: A comparison between agile and traditional software development methods. Technical report, The University of Western Australia (2005)

18. Ulrich, K.T., Eppinger, D.S.: Product Design and Development. McGraw-Hill Education, New York (2011)

19. Inukollu, V.N., Keshamoni, D.D., Kang, T., Inukollu, M.: Factors influencing quality of mobile apps: role of mobile app development life cycle. Int. J. Soft. Eng. Appl. 5(5), 15–34 (2014)

20. Hong, S., Goor, G., Brinkkemper, S.: A comparison of object-oriented analysis and design methodologies. In: 26th Hawaiian Conference on System Sciences, Hawaai (1993)

21. Abrahamsson, P., Salo, O., Ronkainen, J., Warsta, J.: Agile software development methods - Review and analysis. Technical report, Technical Research Centre of Finland (2002)

22. George, B., Williams, L.: A structured experiment of test-driven development. Inf. Softw. Technol. 46(5), 337–342 (2004)

23. Schoeffel, R.: The Concept of Product Usability. ISO Bull. pp. 5–7 (2003)

Cultural Behavior Features for Adapting Hospital Information Systems

Reem Ehaidib and Ajantha Dahanayake[✉]

College for Women, Prince Sultan University,
King Abdullah Road, Riyadh 11585, Saudi Arabia
AjanthaDahanayake@yahoo.com

Abstract. An information system that will be suitable for every user in any circumstances, and could fit to all users' needs and their individuality is not acceptable. Instead of imposing one system for all users' it is most appropriate to consider making systems adaptable to its future users' cultural needs and preferences. Therefore, this research explores the culture specific behavior adaptation in the HIS domain and introduces the cultural behavior features for hospital information systems adaptation.

Keywords: Cultural behavior features · Cultural behavior · HIS · Hospital information systems adaptation

1 Introduction

In the healthcare sector many hospitals opt to invest in successful systems in order to modernize their hospital information systems (HIS) and to manage their patients in a convenient and efficient manner. The unavailability of HIS applications in the local market pushes hospital management's to adopt systems developed by foreign development team. Adopting a system has its disadvantages as the foreign system is implemented based on the requirements that are particular and limited to a foreign hospital situation and their culture and environment. Some of those culture specific needs are not useful to be applied in another location. As this issue costs hospital managements dearly, then they decide to adapt the foreign HIS to their hospital by reconfiguring the system and making it adapted to their hospital's culture and stakeholder behavior.

1.1 HIS Adaptation to KSA Healthcare Industry

A HIS developed by a South Korean (SK) company is being adapted by a hospital in Saudi Arabia. This Korean system provides a vast patient information management functionality and is compatible to the health standards that the Riyadh hospital is targeting to achieve. The SK HIS has been tested in South Korean and is automated to control the overall management in the health care sector. From the onset of this initiative it was evident that the SK system has not considered the system to be customized to the Kingdom of Saudi Arabia (KSA). The SK-HIS system did not consider

© Springer International Publishing Switzerland 2016
M. Ivanović et al. (Eds.): ADBIS 2016, CCIS 637, pp. 180–192, 2016.
DOI: 10.1007/978-3-319-44066-8_19

the compatibility with the Saudi culture. For example, the Saudi religion "Islam" allows men to have more than one spouse and they also did not consider that in all situations female patients need the consent of a male guardian irrespective of the female patient's age.

As a result of these issues the Riyadh hospital's management together with the IT department decided to initiate a research and development plan leading to a more structured and controlled adaptation of the SK system to Riyadh hospital environment. The research presented in this paper belongs to the association agreement of the Riyadh hospital's HIS adaptation project. The research explores the problem of including the stakeholder's cultural behavior during the system's configuration and transformation as one of the problems that needs to be addressed during HIS adaptations. It presents the cultural behavior features that has been identified during a systems adaptation processes.

The cultural features are the things that influence people's individuality, such as religion, spirituality, economic situation, family and community life, and other aspects of users' culture [1]. The term culture according to Hofstede et al. [2] is a collective phenomenon, which is shared with people who live or lived within the same social environment: (a) Culture consists of unwritten rules of the social game; (b) It is the collective programming of the mind that separates the members of one group or category of people from others.

An information system that will be suitable for every user in any circumstances, where the system could fit to all users' needs and satisfy their individuality is not realistic [3]. For example, instead of imposing one system for all users' it is realistic to consider systems adaptability to its future users' cultural needs, behaviors, and preferences. Therefore, this research explores the cultural specific behavior adaptation in the HIS domain and introduces a set of cultural behavior features for HIS adaptation.

The rest of the paper is organized as follows: Sect. 2 provides the background and information on IS and HIS adaptation in general. Section 3 gives an overview of the Riyadh healthcare systems needs analysis. Section 4 provides the identification of the cultural behavior features for HIS adaptation. Section 5 presents the concluding remarks.

2 Systems Adaptation

2.1 Information Systems Adaptation

Information systems adaptation has received relatively little attention in the scientific research even though it has become a common practice [4]. The approach to adapt an already successful system is preferred to building a system from scratch in reality. Software systems adaptation captures the degree to which the system is used deeply, or to its fullest extent, for the improvement of organizational and individual performances [5]. However, it is impossible to anticipate the requirements of all users and potential customers. A best or optimal system configuration is impossible. The active involvement of users and a clear understanding of their tasks and requirements is a challenge during software systems development. It makes even more challenging during systems adaptation to other situations than it is originally made for because [4]: (1) As the potential user groups and potential customers are not known a prior, (2) The requirements that are needed to be identified according to future scenarios are not known. (3) The dynamics of changing conditions shifts the customization process of the system's characteristics from the development phase to its usage and adaptation phase because the time available for professional development in projects are short and the new features are too costly to take into consideration.

For this reason, developers implement techniques of adaptation into the system in order to react to changing conditions as fast as possible [4]. There is an important distinction concerning adaptation techniques and systems adaptation. The adaptation techniques differentiate between manually and automatically performed adaptation processes and decomposes into the two terms adaptivity and adaptability [31].

The systems adaptation in general is the process of changing to fit some purpose or situation, adjustment to environmental conditions as modification of an organism or its parts that makes it more fit for existence under the conditions of its environment [6]. The software systems adaptation is the process of transformation from an old system to a target system of the stakeholder by introducing a system that works successfully for similar functions [7]. The systems adaptation increases the match between the user behavior and systems functionality once the system is adapted to the new customer's environment and the system is elevated to acceptance according to the behavior of its stakeholders [4].

The systems adaptation addresses different varieties along manifold dimensions [4]. The consideration of user preferences like language, color schemes, modality of interaction, menu options or security properties, and numerous other personal preferences are popular sources of adaptation and can be reused in different applications. Other sources are the user's interests and disinterests, psychological personality characteristics like emotions, self-confidence, motivation, or beliefs, which are difficult to assess automatically. Authors in [8] summarize those as the influence of espoused cultural values on user coping strategies of systems adaptation. Moreover, they are frequently required to cooperate spontaneously and opportunistically with previously unknown software services in order to accomplish tasks on behalf of the users.

2.2   Hospital Information Systems Adaptation

The main focus of healthcare industry is customer satisfaction as reported in the Healthcare Information and Management Systems Society (HIMSS) [9]. Therefore, the healthcare industry has pushed to adapt information systems that manage the hospital data and information, widely known as HIS. The hospital market, organizational, and financial factors influence the adaptation of HIS in the healthcare industries [10].

The most prominent example of HIS adaptation to date is the Veterans Information Systems and Technology Architecture (VISTA) [11]. VISTA is a nationwide information system and Electronic Health Record (EHR) developed and deployed by the U.S. Department of Veterans Affairs (VA) throughout the U.S. to all 1200 + healthcare sites of the Veterans Health Administration (VHA), and manages the largest integrated healthcare network in the United States. It provides care to over 8 million veterans, employing 180,000 medical personnel and operating in 163 hospitals, over 800 clinics, and 135 nursing homes throughout the U.S., Alaska, and Hawaii on a single electronic healthcare information network. Nearly half of all U.S. hospitals that have a complete inpatient/outpatient enterprise-wide implementation of an EHR are using VISTA [11]. VISTA software modules have been adapted around the world, and are being considered for adaptation in healthcare institutions such as the World Health Organization (WHO).

In the recent years, the cost of providing high quality services and patient satisfaction in hospitals has increased tremendously. Most hospitals no longer develop their own in-house HIS solutions. They purchase software solutions available in the marketplace. This is problematic for many countries at two levels. First, they do not have the financial resources to acquire a commercial HIS. Second, they do not have the same culture and organization that are implicitly or explicitly imposed and implemented as part of the commercial solutions

Several studies have examined the adaptation of information systems, such as by Richard Heeks [12]. He concludes that the information systems that succeed are those that best incorporate the key technical, social, and organizational environment aspects in which they are implemented. Heeks also noted that the failures are mainly due to the systems adaptation approaches that do not take into account the context, or the local cultural behavior.

3   CASE STUDY: Riyadh HIS

A system developed at an educational hospital in South Korea with the engagement of a Microsoft partner is introduced into the Riyadh hospital, with the goal to offer a highly intuitive user experience and unified communication. The Riyadh hospital is a huge entity that is responsible for providing healthcare services to different parts of Saudi Arabia. The hospital holds more than six thousand beds and maintains more than two million medical records of patients over the last forty years and there are around forty-eight thousand employees working in this hospital.

3.1   Old Systems of the Riyadh Hospital

To maintain the information and data of the patients, the hospital used QuadraMed's Computerized Patient Records (QCPR). QCPR is a legacy system that maintains only the patient's information. In addition, over the years the IT department has developed several applications to help maintaining other healthcare services. The hospital has been using QCPR to manage patient records, the QCPR contains several modules, but the hospital utilized only three main modules: (1) *Registration Module*: responsible for registering new patients to the system. (2) *Admission Module*: responsible for filling patient's information while admitting to the hospital. (3) *Flowchart Module*: responsible for keeping and updating patient's information, plus registering patient's visit and treatments applied to the patient.

The registration module is the only module that is utilized, where no patient will be treated without a profile in the QCPR, regardless of the missing information of some of the patients or the inaccurate information that has been entered to the system. As the QCPR system was bought several years ago, no support or maintenance is provided, it does not have any entry validations, which allows spelling mistakes, redundant information, information to be filled in the wrong fields, and gives the ability to play around with patient information without recording audits. Moreover, the QCPR was responsible in maintaining patient appointments. Unfortunately, the physician or the nurse cannot see the available appointment slots. Only the appointment clerks have the access to this information. This raised issues with the patients who have very late appointments as the physician or the nurse must know about it to extend some of the treatments if needed. In addition, there are inconsistent patient admission information. Each department of the hospital in each region has their own applications that are developed by the in-house IT departments. This led to inconsistency in patient information as some of the many issues and challenges of the old system. This situation has created a lot of redundant work, and most of the healthcare services continued to be managed manually. For this reason, the hospital needed a total system that manages the patient records, the hospital's assets, and the information shared easily with the other regions and healthcare facilities.

3.2   Introduction to Korean HIS

As there is no hospital information system to manage the healthcare services provided to the patients, the management agreed on adopt a HIS that was develop by a South Korean team. The SK HIS is in compliance with Health Level 7 (HL7) [13]. HL7 Internationally defines that the functional model of Electronic Health Records (EHR) has 3 Level-1 categories such as Direct Care, Supportive and Information Infrastructure and those 3 Level-1 categories have 13 detail sub-categories and 230 system functions [27]. In addition, to enhance the patient satisfaction and convenience, the Service Channel is located in front of the entire application. The Service Channel is the application domain for the enhancement of relationship between patient and related organization and it provides services to its customers.

SK HIS is using a typical 3-Tier Architecture (Presentation, Application, and Data) that is most widely applied. All data is saved in Oracle relational database management system (RDMS), and the basic query language is Oracle PL SQL. The application tier using HSF 2.0, which is the middleware hosting all services of SK HIS. It is driven in the version of .NET Framework 4.5.2 or later. All sorts of cross-cutting functions such as Transaction, Data Access, Security, etc. are performed. The presentation tier is developed using Windows Presentation Foundation (WPF) to provide a dynamic user experience.

3.3   Riyadh Hospitals Goals and Needs

– The new adapted SK HIS to Riyadh hospital is meant to manage health care facilities from managing patients from every aspect to managing health care assets.

Riyadh public hospital treats only specific public sector employees and their families[1].

- This system includes several main sections: Service Channel, Direct Care, Supportive Care, and Information Infrastructure and they are provided within the SK HIS.
- In during the implementation phase of the SK HIS system at the Riyadh hospital several differences surfaced and noticed by the implementers and the stakeholders.
- The differences in business rules: HIS system is following SK health standards, whereas Riyadh hospital is following American standards and this has caused major changes in the systems workflows.
- Also cultural differences are visible as the HIS system is develop in SK. The SK and Saudi Arabian cultures are totally different from each other.

The main cultural differences surfaced are:

- *The SK HIS system allows a picture of a female patient to be shown where it is prohibited in the Riyadh hospital.*
- *The SK HIS system allows a female unmarried patient to be treated in the clinics for in-vitro fertilization (IVF). In Saudi Arabia an unmarried female patient cannot be treated in those clinics unless she started the process when she was married and for some reason she is not with her spouse.*
- *In the Riyadh hospital any female patient must have a male guardian and this is not the case in SK.*
- *In the Riyadh hospital the patients are one of the following category: a government employee, a related family member of a government employee (husband, wife, mother, father, son, or daughter), or a royal family member and their treatments free of charge, unless the patient is not in the above categories and is in need of critical care in the emergency room. The hospital is obligated by the government to treat those patients. This is not the case in SK hospital; there the patients are charged for their treatments.*
- *In the Riyadh hospital if a registered patient is under 18 years, male guardian's information is needed, who will be responsible for approving any procedures. Whereas, in the SK hospitals all female or male patients under 18 years needs either female or male guardian.*
- *In the Riyadh hospital all female patients need to provide male guardian's information and he will be responsible for approving any procedures. Whereas in the SK hospital a female patient above 18 years is responsible for approving her own procedures.*
- *In the Riyadh hospital a married female patient needs to provide her husband's information, where a married male patient does not have to provide his wife(s) information and an emergency contact is sufficient. Whereas in the SK hospital a married patient needs to provide their spouse's information regardless of the patient is male or female. The Riyadh hospital faces problems in controlling*

---

[1] For confidentiality reasons the specified public sector cannot be mentioned.

infectious disease spread due to the lack of information in case a male patient catches such a disease.

- *In the Riyadh hospital there are different levels (eligibility type) of patients: regular patients, VIP patients, and Royal patients and for each type the characteristics of the provided hospital rooms will be different. The Royal patients have their own physicians and nursing staff who are allowed to enter their wards and allowed to see their medical reports and charts. Whereas in the SK hospitals the patients are of the same level. All patients are charged for their treatments so they do not need to differentiate.*

Those differences have cost both parties, time, effort, and money to maintain, translate, adapt and implement the SK HIS at the Riyadh hospital. Therefore, the IT department of Riyadh hospital decided to take over the HIS systems adaptation initiative.

3.4   Mapping: Riyadh HIS to Korean HIS Functional Areas

The Riyadh HIS system model is divided into four main domains: Service Channel, Direct Care, Supportive Care, and Information Infrastructure. The Riyadh HIS system has 33 modules in total distributed within the four domains. The detailed Riyadh HIS modules are given in Fig. 1.

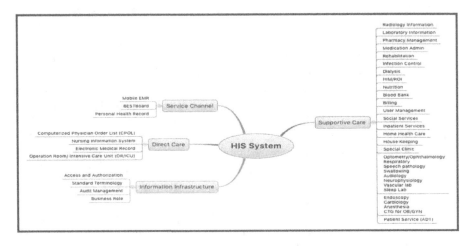

Fig. 1. HIS system modules

During the needs analysis of the SK HIS adaption to Saudi Arabia has shown strong culture differences of the user's behavior compared to SK. The cultural challenges represented in this system are mainly religion, nationality, and gender. In addition, the conflict of the services provided to patient with those cultural differences decrease the user satisfaction. As the first phase only the following 4 modules are considered for the systems adaptation imitative. Therefore, the 4 main modules that is covered in this research are:

- *Electronic Medical Record, Nutrition, User Management, Patient Services*

4   The Cultural Behavior Features

Hofstede [14] defines the culture as a "the collective programming of the mind that distinguishes the members of one category of people from another". The culture is learned from childhood or acquired by being affected from other cultures; it plays an effective role in most life aspects [15]. Hofstede believe that the mental activities will be affected by the culture [16]. In the cultural field of research Hofstede and Lewis theories are considered important. The Hofstede's theory [26] considers five dimensions of the culture: power distance (PD), individualism (IDV), masculinity (MAS), uncertainty avoidance (UAI), and long term orientation (LTO). Lewis [17] categorizes the cultures into three classifications: linear-active, multi-active, and reactive. Further, culture as a multidimensional and dynamic concept a user model for Multicultural Adaptive Systems is presented in [18].

Reinecke and Bernstein [19] present the influence of adapting the cultural factors into user interfaces (UI) and test their theory by developing a prototype of a web application. The authors discuss and use Hofstede's theory [16]. Also, they found the differences between the multicultural user interfaces and user interface of particular culture. They evaluate the adaptation rules of MOCCA[2] in order to configure suitable user interface for each specific user, specific to his/her national culture. The important part of this study is the cultural factors that are collected from other researches and the detailed discussion.

Eristi [20] has studied the cultural factors which are determined through 15 university sites from 11 different countries. She describes Marcus and Gould [21] theory which explains the effect of Hofstede's theory on the interface designs. The author discusses the cultural characteristics that affect the web design. In addition, Breitenöder's [22] study the cultural characteristics that affect the web design for marketing purposes. Some of the collected cultural factors from several researches in [23] give consideration to the political orientation and social structure. In addition, some of the factors' are considering the design activities for privacy considerations, such as nationality of the father and mother [23]. Table 1 summarizes the cultural factors and their related fields that are discussed in several articles.

Table 1. Summary of cultural factors that influence UI functionality

Main Cultural Factors	Related fields	References
Language	First language	Nisbett [24], Reinecke and Bernstein [19]
	Second language	
Reading/writing direction	Left-to-right	Chan and Bergen [25], Reinecke and Bernstein [19], Röse [26]
	Right-to-left	
	Right-to-left/top-to bottom	

(*Continued*)

---

[2] MOCCA is a name of culturally adaptive system that is introduced by authors in [19].

Table 1. (*Continued*)

Main Cultural Factors	Related fields	References
Religion	Religion	Sialaa, et al. [27], Eristi [20], Breitenöder [22]
	Symbols	
	Color	
Education level	Education level	Microsoft [28], Reinecke and Bernstein [19]
Form of instruction	Computer literacy	Liegle and Janicki [29], Reinecke and Bernstein [19]
Personality	Gender	Reinecke and Bernstein [19], Kamentz and Womser-Hacker [30]
	Date of birth	
Location	Former residence	Reinecke and Bernstein [19], Breitenöder [22]
	Former residence duration	
	Current residence	
	Current residence duration	
Nationality	Nationality	Reinecke and Bernstein [19]
Visual information	Sound/music	Eristi [20]
	Videos	
	Visual images	
Aesthetic factors	Using aesthetic elements/or more abstract	Eristi [20]
	Design elements, Design principles (such as length of paragraph or numbers of words)	
Non Functional factors	Non-functional requirements	Eristi [20]
	Resolution, Size, Measure, Loading pace	

Table 2. Cultural Features Assessment across the HIS main modules (UM: User Management, ERM: Electronic Medical Record, N: Nutrition, PS: Patient Services)

Main Cultural Features	Requirements	UM	EMR	N	PS
Language	First language	✓	✓	✓	✓
	Second language	✓	✓	✓	✓
Reading/ writing direction	Left-to-right	✓	✓	✓	✓
	Right-to-left	✓	✓	✓	✓
Religion	Religion	✓		✓	✓

(*Continued*)

Table 2. (*Continued*)

Main Cultural Features	Requirements	UM	EMR	N	PS
Form of instruction	Technology background and the ability of using computer				
Personality	Employee Badge	✓	✓		
	Emergency contact information	✓	✓		
	Gender	✓	✓	✓	✓
	Date of birth	✓	✓	✓	✓
	Marital status	✓	✓		✓
	Contact information	✓	✓		
	Patient eligibility type	✓	✓		✓
	Identification ID	✓	✓		
Location	Current residence	✓	✓	✓	
	Current residence duration	✓	✓	✓	
Nationality	Nationality	✓	✓	✓	
Visual information	Sound/music	✓	✓		
	Videos	✓	✓		
	Visual images	✓	✓	✓	✓
UI setting	Using aesthetic elements/or more abstract	✓	✓		
	Design elements, Design principles	✓	✓	✓	✓
(Non-Functional) UI Quality	Resolution, Size, Measure, Loading pace	✓	✓	✓	✓
	Reliability (*Fault Tolerance, Recoverability, Accuracy, Security*)	✓	✓	✓	✓
	Usability (*Understandability, Learnability, Attractiveness, Suitability*)	✓	✓	✓	✓
	Efficiency	✓	✓	✓	✓
	Maintainability	✓	✓	✓	✓
	Portability	✓	✓	✓	✓
Date/Time	Calendar (*Hijri/Gregorian*)	✓	✓	✓	✓
	Time Zone	✓	✓	✓	✓

The cultural factors that influence the UI behavior in Table 1 are integrated with the Riyadh HIS service requirements (Sect. 3: CASE) to arrive at the list of cultural features (Table 2). This list of cultural features is then mapped to the 4 Riyadh HIS modules used in this research to assess the validity of cultural features that are necessary for the identification of stakeholder's cultural behavior features.

Culture as a multidimensional and dynamic concept as presented in the article of Thalheim and Jaakkola, "Multicultural Adaptive System" [18] has proposed an approach to incorporate culture into information systems, by classifying the stakeholders' culture into stereotypes. They have presented mainly two stereotypes: user portfolio and user profile, which combine the tasks, involvement of users, collaboration, restriction, education profile, work profile, personality profile and security profile.

Our focus in this research is on those two stereotypes: user portfolio and user profile for abstracting the stakeholder or user behavior. The user profiles are types of user roles mainly differentiating the systems users and patients. They perform within HIS services according to the user's role (e.g. Systems users such as Registration clerk, nutrition nurse, surgeon, physician, radiologist, etc. and the patient types such as Royal, Normal citizens, foreign/temporary residents). The user portfolio defines the tasks to be performed by the users. Then the features that influence the user interface (UI) can be categorized according to: systems quality, systems setting, data inputs.

Therefore, the Stakeholders' cultural behavior features are derived as follows:

User behavior stereo types are: portfolio and profile

User Profile defines the types of user roles and provides services to: Systems users and Patients.

– Systems users are: registration clerk, nutrition, ward management, etc....
– Patient types are: Royal, Citizens, and Temporary Residents

User Portfolio defines the tasks to be performed for: UI Quality, UI settings, and UI Data-input

Data-inputs are of two categories: standard data and cultural behavior data

Standard data are those that are already available in the main system such as:

– First Language, Emergency Contact Information, Gender, Date of birth, Contact Information, Identification ID, Current Residence, Nationality,

Cultural behavior data are those need to be included during the systems adaptation:

– Second Language, Reading/Writing Direction, Religion, Patient Type, Marital Status, Patient Eligibility, Duration at Current Residence, Nationality, Sound/Music, Videos, Visual Images, Calendar type, Time Zone.

UI Quality and UI Settings are general features of any system and they cannot be considered as cultural behavior features during the HIS adaptation process.

5   Concluding Remarks

The main contribution of this research is the identified cultural behavior features: Second Language, Reading/Writing Direction, Religion, Patient Type, Marital Status, Patient Eligibility, Duration at Current Residence, Nationality, Sound/Music, Videos, Visual Images, Calendar type, Time Zone.

Those features are used during the adaptation of the 4 modules of SK HIS user interface to Riyadh HIS user interfaces. The cultural features helped the developers in finding exact points where cultural features need to be included. The IT team of the Riyadh hospital successfully completed the adaptation of user interfaces according to the needed cultural behaviors of Riyadh hospital stakeholders for those 4 modules. Based on the positive input of developers at the IT unit and success of adapted UI's of the 4 modules the hospital management decided to continue with the adaptation of rest of the Riyadh HIS modules using the cultural features developed in this research.

References

1. Cultural (2015): http://www.Merriam-Webster.Com/Dictionary/Cultural. Accessed on 29 April 2015
2. Hofstede, G., Hofstede, G.J., Minkov, M.: Cultures and organizations: software of the mind: intercultural cooperation and its importance for survival, 3rd edn. McGraw-Hill, New York (2010)
3. Geambaşu, C.V., Jianu, I., Jianu, I., Gavrilă, A.: Influence factors for the choice of a software development methodology. Acc. Manage. Inf. Syst./Contabilitate Si Informatica De Gestiune 10(4), 479–494 (2011)
4. Elie-Dit-Cosaque, C.: Studies on adaptation to information systems: multiple roles and coping strategies. PhD Dissertation, Georgia State University USA (2009)
5. Fadel, J.K.: User adaptation and infusion of information systems. J. Comput. Inf. Syst. 52(2), 1–10 (2012)
6. Adaptation (2016): http://www.merriam-webster.com/dictionary/adaptation. Accessed on 14 February 2016
7. Eason, K.D.: Information Technology and Organizational Change. CRC Press, Boca Raton (2005)
8. Srite, M., Karahanna, E.: The role of espoused national cultural values in technology acceptance. MIS Q. 30(3), 679–704 (2006)
9. Healthcare information and management systems society. In: 19th Annual HIMSS Leadership Survey Sponsored By Cisco Final Report (Rep.). Healthcare CIO (2008) http://www.Himss.Org/2008Survey/DOCS/19thannualleadershipsurveyfinal.pdf
10. Wang, B.B., Wan, T.T., Burke, D.E., Bazzoli, G.J., Lin, B.Y.: factors influencing health information system adoption in American hospitals. Health Care Manage. Rev. 30(1), 44–51 (2005)
11. Brown, S.H.: VISTA, U.S. Department of Veterans Affairs national scale HIS. Int. J. Med. Informatics 69(2–3), 135 (2003)
12. Heeks, R.: Health information systems: failure, success and improvisation. Int. J. Med. Inf. 75(2), 125–137 (2006)
13. Health Level Seven International–Homepage (2015). http://www.HL7.Org/. Accessed on 24 October 2015
14. Hofstede, G.: Cultures and organizations: Software of the mind. McGraw-Hill, London (1991)
15. Corsaro, W.A.: Interpretive reproduction in children's play. Am. J. Play 4(4), 488–504 (2012)
16. Hofstede, G.: Culture's consequences: Comparing values, behaviors, institutions, and organizations across nations, second edition, 2nd edn. Sage Publications, Thousand Oaks (2001)
17. Lewis, R.D.: When cultures collide: leading across cultures. Boston Nicholas Brealey International, London (2006)
18. Jaakkola, H., Thalheim, B.: Adoptive system for multicultural deployment. In: 24th International Conference On Information Modelling And Knowledge Bases, pp. 210–229 (2014)
19. Reinecke, R., Bernstein, A.: Knowing what a user likes: a design science approach to interfaces that automatically adapt to culture. MIS Q. 37, 427–453 (2013)
20. Eristi, S.D.B.: Cultural factors in web design. J. Theor. Appl. Inf. Technol. 9(2), 117–132 (2009)

21. Marcus, A., Gould, E.W.: Crosscurrents: cultural dimensions and global web user-interface design. Interactions **7**(4), 32–46 (2000)
22. Breitenöder, A.F.: The Impact of Cultural Characteristics on International Web Design in Marketing Communications. Diplomica Verlag GmbH, Hamburg (2007)
23. Schmid-Isler, S.: The language of digital genres-a semiotic investigation of style and iconology on the world wide web, in System Sciences. In: Proceedings of the 33rd Annual Hawaii International Conference, Los Alamitos (200)
24. Nisbett, R.E.: The geography of thought: Asian and Western minds at work. Free Press, New York (2003)
25. Chan, T.T., Bergen, B.: Writing direction influences spatial cognition. In: Proceedings of the 27th Annual Conference of the Cognitive Science Society, Stresa (2005)
26. Röse, K.: Aspekte der interkulturellen Systemgestaltung. In: Mensch & Computer 2005, Workshop-Proceedings der 5. fachübergreifenden Konferenz, Linz (2005)
27. Sialaa, H., O'Keefeb, R.M., Honea, K.S.: The impact of religious affiliation on trust in the context of electronic commerce. Interact. Comput. **16**(1), 7–27 (2004)
28. Microsoft: Accessible Technology in Computing–Examining Awareness, Use, and Future Potential (2004)
29. Liegle, J.O., Janicki, T.N.: The effect of learning styles on the navigation needs of web-based learners. Comput. Hum. Behav. **22**(5), 885–898 (2006)
30. Kamentz, E., Womser-Hacker, C.: Lerntheorie und Kultur: eine Voruntersuchung für die Entwicklung von Lernsystemen für internationale Zielgruppen. In: Proceedings of Mensch und Computer: Interaktion in Bewegung, Stuttgart (2003)
31. de Lope, J., Maravall, D.: Adaptation, anticipation and rationality in natural and artificial systems: computational paradigms mimicking nature. Nat. Comput. **8**(4), 757–775 (2009)

An Implementation Method of an Information Credibility Calculation System for Emergency Such as Natural Disasters

Ken Honda[1]([✉]), Naoki Ishibashi[2], and Naofumi Yoshida[2]

[1] Graduate School of Global Media, Komazawa University, Tokyo, Japan
Ken.h.0802@gmail.com
[2] Faculty of Global Media Studies, Komazawa University, Tokyo, Japan

Abstract. In this paper, an implementation method of an information credibility calculation system is introduced. This system objectively calculates information credibility by comparing target information with various information resources on World Wide Web and sensor data. Information credibility is calculated by the degree of objectivity. This paper shows the feasibility of the method by experiments.

Keywords: Credibility · Objectivity · Web · Sensor data

1 Introduction

The Great East Japan Earthquake that occurred in March 11th 2011 had paralyzed many information functions and confused many people. After that, some disaster such as landslide and eruption were continuing. There is possibility that occur earthquakes all around the world or eruptions of large mountains. Thus tools to solve the situation become more important. The ideal state of ICT and media is working anytime and providing important information 24 h, 7 days. The information system has to work even in natural disasters. If there were helpful system, it would be available for collecting information when normal period without emergency.

There are various harms caused by uncertain information. For example, in the Great East Japan Earthquake, rumor information about relief supplies was widely spread through chain e-mails or SNS (Social Networking Services). For example that is "Iwate prefecture needs relief supplies from an individual". This was not true. Eventually the public relations department of Iwate prefecture local government officially denied it [15, 18]. Spreading rumor information with the lack of accurate information was obstacle, it caused wastes of relief supplies and increase anxieties among people.

Based on these backgrounds, this paper introduces an implementation method of the system that calculates information credibility. The information credibility is calculated by the degree of objectivity. The degree of objectivity is detected by referring various information resources and sensor data in the method (Fig. 1). There have been a lot of studies about information credibility. In the paper [11], the method analyzes relation between information. Especially, similarity and co-occurrence relation (support) between target information and related information of it have been analyzed.

M. Ivanović et al. (Eds.): ADBIS 2016, CCIS 637, pp. 193–201, 2016.
DOI: 10.1007/978-3-319-44066-8_20

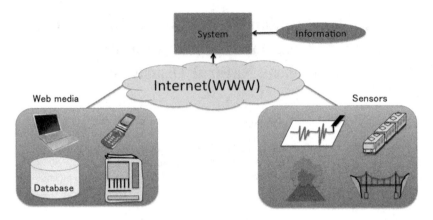

Fig. 1. Summary of system

Information that has many supports is high credibility. This research calculates information credibility by similarity and co-occurred relations among media. In the paper [9], the method has analyzed SNS to prevent extract unsuitable information. It has analyzed expression in a sentence and elements of tweet. In the paper [6], it analyzes the tendency of rumor information, revising rumor information and senders themselves based on six hypotheses on micro blog.

This paper proposes an implementation method of system that can support to collect information in case of emergency such as natural disasters. This method calculates information credibility by comparing target information with various information resources on World Wide Web and sensor data.

The credibility is not a property of an object, person or piece of information but is "a perceived quality" [4]. The credibility calculation is included the trust management.

In the paper [5], trust management techniques are categorized in four types. They are: policy as a trust management technique, recommendation as a trust management technique, reputation as a trust management technique, and, prediction as a trust management technique. One of the important issues of credibility is the source credibility [6]. Major techniques of text based credibility calculation methods use machine learning (for example [7]). Our method is recommendation technique of the credibility (recommending the degree of credibility), focusing source credibility, and not using machine learning, using sensor data as source of the trust.

Contributions of this method are the following two points. The first one is the utilization of several reliable information resources. The second is the utilization of objective sensor data. This method integrates these two points. By using several resources and sensor data, the method is able to derive the value of objectivity. We consider that is information credibility.

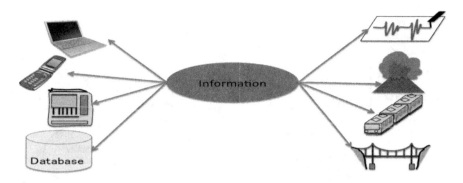

Fig. 2. Comparing target information with web, sensor data

2 Concept of the Method

In this method, we have used both several reliable information resources and sensor data that are opened on the web. The method calculates target information credibility by comparing those information. Figure 2 shows the overview of this method.

In specifically, reliable information resources are government institution (Cabinet, Ministry, the Self-defense Forces, Police, Fire Agency), press relations (National and local newspaper, Press agency, Company) and volunteer groups (NPO, NGO, supporting organizations). And sensors are observation sensors (earthquake, volcano, storm), disaster observe satellites, bridge sensor and railroad sensors.

Figure 3 shows an overview of the method that calculates inputted target information credibility. First, this system derives real time data and store to database in this system from reliable resources. Second, input target information to database as a text. Third, this system compares target information with resources. Finally, the system calculates ratio of matching and outputs the value from 0 % to 100 %. Then, we are able to get information credibility by evaluating the numerical value.

As a result, the value proves degree of objectivity for target information. It will be the degree of conclusive credibility.

3 Credibility Calculate System

In this section, the method of matching between target information and several information resources is shown.

First, we explain about method to match with several information resources and target information. In a current implementation, this system collects real time information constantly by using HTML parser from websites of several resources. The information was stored to database per 1 sentence. For example, if the sentence is "an earthquake is occurred in Tokyo". Then, the system splits the sentence (Fig. 4).

Next, users input target information into the system. The target information also split into words. The word matching is performed in brute force manner for resource

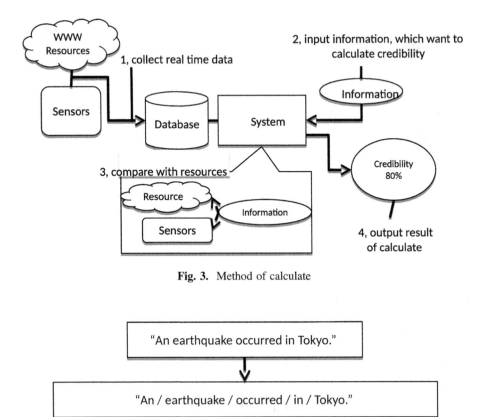

Fig. 3. Method of calculate

"An earthquake occurred in Tokyo."

"An / earthquake / occurred / in / Tokyo."

Fig. 4. Split the sentence

information and target information. The process is repeated only the number of sentence in database (Fig. 5).

To calculate a numerical value indicating information credibility, the total matching number are divided by a value that is obtained by multiplying the total character number of the target information and the total character number of resources in database. This numerical value shows the target information credibility. Figure 6 shows it in the numerical formula. "n" shows the total matching number. "length_a" shows the total character number of the target information, and "length_b" shows the total character of the various information resources.

Second, we explain about the method to use sensor data. Sensor and method suitable for it are various. As a sample, we use "Strong Motion Seismograph Network" provided by NIED (National Research Institute for Earth Science and Disaster Resilience). This sensor indicates earthquake locations, time and size in Japan in real time and shows them by each color in the map. It updates earthquake situation every two seconds. Not magnitude but peak ground acceleration is used in this sensor. For example, color "red" shows more than 100gal that is equal to intensity of more than 5. That is, "red" shows a huge earthquake. Earthquakes less powerful than "red"

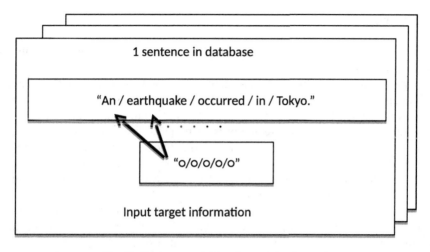

Fig. 5. Sentence matching (present implement)

$$\text{Credibility} = \frac{n}{(\text{length\_a} \times \text{length\_b})}$$

Fig. 6. Calculation of credibility

including slight shaking or shaking related to daily life, are also shown by each color. Information credibility about earthquake itself can be calculated from it. Regarding the usage of this sensor, the system focused on "red". The system captures a map data only when "red" is shown in the map. The occurrence of earthquake is detected as 0 or 1 for the value of Sensor.

The system calculates information credibility by using the two numerical values indicating information credibility that is obtained in each ways. One is the value that is obtained from analysis of the matching number between the target information and various reliable information resources. Another is the value that is detected from sensor data. By integrating these values, the system is able to calculate the value of information credibility more precisely. Finally, the system summarizes these two values (the credibility and sensor data) shown in Fig. 7.

$$FinalCredibility = \begin{cases} 0 \,(where\ Sensor = 0) \\ \dfrac{n}{length_a \times Length_b} \,(where\ Sensor = 1) \end{cases}$$

Fig. 7. Calculation of final credibility ratio

4 Experiment

In this section, the feasibility of implementation method is introduced especially for calculation of information credibility.

In the system that uses web information, it extracts real time data from the Guardian "natural disaster and extreme weather", headline of Euro News and headline of Czech News as a data set. In advance, the system stores this information in database. The target information to calculate credibility is "earthquake in Texas" (No.1), "the thief group appears" (No.2), "shark incident near Perth" (No.3). No.1 is information that is only whether earthquake occurs. No.2 is not stored in database. No.3 is included only one having high degree of match in database. Above information are entered in database and used in the experiments.

First experiment is carried out used an input information No.1 "earthquake in Texas". Table 1 is the result of first experiment. The "db_text" is information from resources in database. The "ratio" is numerical value of matching with input information. The "sensor" is the numerical value of sensor data. If the earthquake occurred, the value would be 1. In Table 1, the "sensor" is 0.

Table 1. Sensor data is 0

db_text	ratio	sensor
Texas floods: Gulf coast emergency worsens after soldiers' deaths Published: 9:56 PM	0.0006	0
Lightning strikes injure scores at Germany's Rock am Ring festival Published: 5:50 PM	0	0
France floods: third person dies as river levels begin to fall Published: 12:39 PM	0	0
Wild weather wreaks havoc on Australia's east coast Published: 8:38 AM	0	0
Paris floods: 'There's something terrifying about it' Published: 7:26 AM	0	0
Texas floods: bodies of four soldiers found as death toll rises Published: 3:08 AM	0.0006	0

Two information matches with "Texas". In present implement, multiply ratio and sensor data together. Therefore, the credibility of input no.1 is zero. We cannot judge whether the information credibility. If the "sensor" were 1, the ratio would not change. Thus, the "ratio" is numerical value of credibility.

Integrating sensor data and web data, it can be help to judge information credibility. By using sensor, objectivity becomes high. Therefore, information credibility can be calculated more precision.

Table 2. There is no information to matching

db_text	ratio
Diver' s death treated as second 'fatal shark incident' near Perth in a week	0
Swiss advocates of basic income for all claim moral victory, despite crushing defeat	0
Thousands rally in Polish capital against conservative government	0
03.06.2016, 21:22 Nuke plant Temelin started planned shutdown of unit 2, stopped producing electricity at 21:00. Some one quarter of fuel will be replaced at the unit.	0.0002
05.06.2016, 09:27 Czech embassy in France conformed incident in which Czech bus was hit with gunshots in the south of country. It was probably random shooting.	0.0002
03.06.2016, 14:51 Czech President Milos Zeman will visit Armenian genocide museum during his visit to Armenia next week. He will lay wreath and plant tree there.	0.0002

Table 3. There is the matching data in database

db_text	ratio
Diver' s death treated as second 'fatal shark incident' near Perth in a week	0.0019
Swiss advocates of basic income for all claim moral victory, despite crushing defeat	0
Thousands rally in Polish capital against conservative government	0
03.06.2016, 21:22 Nuke plant Temelin started planned shutdown of unit 2, stopped producing electricity at 21:00. Some one quarter of fuel will be replaced at the unit.	0
05.06.2016, 09:27 Czech embassy in France conformed incident in which Czech bus was hit with gunshots in the south of country. It was probably random shooting.	0
03.06.2016, 14:51 Czech President Milos Zeman will visit Armenian genocide museum during his visit to Armenia next week. He will lay wreath and plant tree there.	0

Next experiment is carried out used an input information no.2 "the thief group appears" (Table 2). In this experiment, there is no match information in database. Matching ratio is 0 or extremely low. The sensor, which used this time, is earthquake sensor. Therefore, in the experiment no.2 and 3, values in the columns "ratio" are calculated as *FinalCredibility* shown in Fig. 7.

When the input only matches with "the", ratio is not 0.

Final experiment is carried out used an input no.3 "shark incident near Perth" (Table 3). In this experiment, input information is similar to only one information in database.

The ratio is almost 0 or low. However, if the input information matches completely, the ratio is higher than each other.

This experiment shows that credibility can be calculated by deriving objectivity by this method. If there were similar sentence to input sentence in database, each ratio would be high. Therefore, integrating with sensor is required and important. Also, in database, if there were the information that revises past information, the ratio became high. However, integrating sensor can correspond this problem.

This experiment shows that this system is able to match accurate and the importance of integrates for information with diversity. It shows usability of sensor data to derive information objectivity.

5   Conclusion and Future Work

In this paper, we have introduced that calculation of information credibility on World Wide Web especially in case of emergency such as natural disaster by using sensor data. Furthermore, the feasibility of this system is shown by the experiments. During disasters, people have difficulty to judge things correctly. In this situation, one of the serious problems is rumor. People tend to be misled by rumors during disasters. The purpose of this method is to derive information objectivity to help people when they collect information. In this background, this research aims for realizing the information credibility calculation system as an indicator for information gathering. Therefore, we have introduced the method that use several resources and the method that use the sensor data to gain extremely objective information.

However, we still need more considerations. The method needs to adopt machine learning, morphological analysis, Word2Vec neural networks to matching more accurate with text to resource. And it also needs to obtain appropriate sensor and the approach of matching. Furthermore, we need to contrive and adopt the other method. For example, information importance changes by temporal weighting for web and sensor information, and matching to event occurrence time and input information time.

Acknowledgement. We have greatly benefit from Strong Motion Seismograph Network provided by National Research Institute for Earth Science and Disaster Resilience. This work was supported by JSPS Grants-in-Aid for Scientific Research KAKENHI Grant Number 26330141.

References

1. Flanagin, A.J., Metzger, M.J.: The credibility of volunteered geographic information. GeoJournal **72**, 137–148 (2008)
2. Salganik, M.J., Dodds, P.S., Watts, D.J.: Experimental study of inequality and unpredictability in an artificial cultural market. Science **311**, 854–856 (2006)

3. Young, S., Hilligoss, B.: College student's credibility judgements in the information-seeking process, pp. 49–72. The MIT press, Cambridge (2008)

4. Tseng, S., Fogg, B.J.: Credibility and computing technology. Commun. ACM **42**(5), 39–44 (1999)

5. Noor, T.H., Sheng, Q.Z., Zeadally, S., Jian, Y.: Trust management of services in cloud environments: obstacles and solutions. ACM Comput. Surv. **46**(1), 30 (2013). Article 12

6. Pornpitakpan, C.: The persuasiveness of source credibility: a critical review of five decades' evidence. J. Appl. Soc. Psychol. **34**(2), 243–281 (2004)

7. Wawer, A., Nielek, R., Wierzbicki, A.: Predicting webpage credibility using linguistic features. In: Proceedings of the 23rd International Conference on World Wide Web, pp. 1135–1140 (2014)

8. Umejima, A., Miyabe, M., Aramaki, E., Nadamoto, A.: Tendency of rumor and correction re-tweet on the twitter during disasters. IPSJ SIG Technology Report, vol. 2011-DBS-152, No.4, vol. 2011-IFAT-103, No.4. Information Processing Society of Japan (2011)

9. Uchino, T.: Reliability Assessment and Analysis in the Informal Retweet of Twitter. University Faculty of Software and Information Science, Iwate Prefecture (2013)

10. Hasegawa, Y., Oguchi, M.: An access control method which depends on the situation based on a judgment of emergency. DEIM Forum (2013)

11. Mizuno, J., Kruengkrai, C., Ohtake, K., Hashimoto, C., Torisawa, K., Kloetzer, J.: Performance evaluation of disaster information analysis system DISAANA and its question answering mode(in Japanese). IPSJ SIG Technical Report, vol. 2015-MBL-76 No, 14 vol. 2015-CDS-14 No, 14 (2015)

12. Yamagishi, Y., Saitou, K., Mutou, N.: A comparison of video ranking methods based on a time series analysis of tags using multi-category order statistic. DEIM Forum (2015)

13. Yamamoto, Y., Tanaka, K.: Modeling and measuring web information credibility by analyzing support relation between data-pairs. Database **3**(2), 61–79 (2010). Information Processing Society of Japan

14. Yamamoto, Y., Tanaka, K., enhancing credibility judgement of website. WebDB Forum 2010 (2010)

15. Davidson, D.: Subjective, Intersubjective, Objective. Oxford University Press Inc, New York (2007)

16. Surowiecki, J.: The Wisdom of Crowds. Anchor, New York (2009)

17. Ogiue, C.: Inspect, Rumors in the Great East Earthquake(Japanese). Kobunsha Co. Ltd., Tokyo (2011)

18. Cabinet Office. Government of Japan HP, http://www.cao.go.jp

19. Iwate Prefecture HP. http://www.pref.iwate.jp/index.html

20. Iwate prefecture official Twitter. https://twitter.com/pref_iwate/status/46891690585370625

21. Ministry of Land, Infrastructure, Transport and Tourism HP. http://www.mlit.go.jp

Model Capsules for Research
and Engineering Networks

Bernhard Thalheim$^{(\boxtimes)}$ and Marina Tropmann-Frick

Department of Computer Science, Christian-Albrechts-University Kiel,
Olshausenstrasse 40, 24098 Kiel, Germany
thalheim@is.informatik.uni-kiel.de

Abstract. Multi-model utilisation is a common practice in many sciences, e.g. computer science. Coherence and co-evolution of models is however still an open problem. Multi-model approaches suffer however from the impedance mismatch due to differences in modelling languages. The collaboration approach is based on preservation of local models and on explicit association of derived sub-models. Each discipline has developed its specific know-how in modelling and model deployment. Models evolve in dependence on the progress of the research work. If a model or one of its sub-models has been exchanged with a team member then this evolution must also be applied to models of the partner if those sub-models are used elsewhere.

We develop a novel approach to multi-model development and utilisation, to common use and utilisation of models and modelling experience, to systematic assessment of models and systematic extraction of the potential and capacity of models for a research community, and to the co-evolution of model networks.

1 Introduction

1.1 Complex Problems are Solved in Interdisciplinary Communities

Consider two typical research situations and problems that can be observed for interdisciplinary research in *interacting communities* of researchers.

(1) What are the causes for an inflammatory disease, especially for those triggered by dysfunction at boundary surfaces? Why is it a phenomenon of civilisation? How are cells and tissues infected? What kind of patient-specific treatment can be developed? How can life with a disease be improved and under what circumstances? What societal changes are required to move towards preventive medicine? Many branches of biology, medical science, economy, and social sciences participate in the research team (e.g. the Cluster of Excellence "Inflammation@Interfaces" at CAU Kiel). The collaboration relies on models that are exchanged within the team and that are the basis for a common understanding. The use of models is different. Such teams typically span over all four facets of scientific methods. Models are used in

© Springer International Publishing Switzerland 2016
M. Ivanović et al. (Eds.): ADBIS 2016, CCIS 637, pp. 202–214, 2016.
DOI: 10.1007/978-3-319-44066-8_21

the way how empirical sciences use them, e.g. for exploration, experimentation, interpretation, and hypothesis exploration. At the same time models are used in the setting of theory-oriented sciences for explanations, for exploration, for illustration, for proofs and for concept surveys. In computational science models are used for instance for simulation, for emulation of complex processes, for refinement of a general model by data, and for prognosis. Data sciences use models for detection of pattern, for mining of relations, and for generation of hypotheses. Models are the main exchange instrument for scientists in such teams.

(2) How climate is going to change in the future? How much will this change affect daily life? How should society and politics respond today? In order to answer such questions, teams with different backgrounds and from different sciences must be brought together. Teams must have in-depth expertise in their specific area, a common understanding, and a culture of collaboration. Each discipline and team member uses a specific background, a specific way of working, a specific language and a specific manage data and information. Team members need to exchange their insights and knowledge through models if they are to be easily understood and integrated in a multi-disciplinary manner. Reliable judgements would have to be made for example for climate change forecasting. To this date, this collaboration is not satisfactorily supported; resolving this issue will be a major research breakthrough.

The same situation can also be observed in Computer Engineering. Large systems typically consist of several components. They are developed in teams where a team member solves a certain development task with a specific scope and with an appropriate model. For instance, UML proposes several dozens of diagram languages for system development, e.g. use case, class, object, activity, package, interaction, sequence, time diagrams. Models developed vary in their scopes, aspects and facets they represent and their abstraction. Multi-modelling [3,11,20,23,24] is a culture in computer science. Maintenance of coherence, co-evolution, and consistency among models has become a bottleneck in development.

1.2   Multi-modelling is the State-of-the-Art in Research and Engineering

Disciplines often use a combination of empirical research that mainly describes natural phenomena, of theory-oriented research that develops concept worlds, of computational research that simulates complex phenomena and of data exploration research that unifies theory, experiment, and simulation [10]. All these research methods use models as one of their main instruments. Typically, a suite or ensemble of models is simultaneously used due to the complexity of the real world, due to orientation on some of the aspects and facets, due to the abstraction level that fits best to the investigation goal, and due to the supporting instruments such as mathematics and visualisation.

Most disciplines integrate a variety of models or a *society of models*, e.g. [2, 14]. Models used in computer science are mainly at the same level of abstraction.

It is already well-known for threescore years that they form a *model ensemble* (e.g. [8,21]) or *horizontal model suite* (e.g. [3,26]).

One of the main obstacles beside coherence of models is co-evolution of models within a model suite. However, this can be supported by strict or eager binding with some toleration of deviation. Coherence can be based on collaboration modi such as master-slave or handshake protocols. It is however an unsolved problem how shared elements can be managed within a model suite. At present, models are in some kind of coopetition (cooperation and competition) within a model suite. Often different languages, different backgrounds and different modelling styles are used and are not harmonised.

1.3   Overview of the Approach

In this paper we tackle the collaboration challenge by developing a flexible system to manage locally and to exchange globally models for collaboration in networks. In this case, models become thus a crosscutting concern to reflect competence for an interdisciplinary collaboration and for interactive research on complex society issues that cannot be solved within a singleton discipline.

We remind in Sect. 2 a novel notion of the model. This notion generalises the notions of models used in archeology, arts, biology, chemistry, computer science, economics, electrotechnics, environmental sciences, farming and agriculture, geosciences, historical sciences, humanities, languages and semiotics, mathematics, medicine, ocean sciences, pedagogical science, philosophy, physics, political sciences, sociology, and sport science.

Next we discuss in Sect. 3 model-based collaboration in research and development. This research can be supported by model suites. They establish coherence maintenance among models. The main novel contributions of the paper are the introduction of the notion of the   **model capsule** in Sect. 4 and the proof of concept in Sect. 5.

2   The Notion of the Model

Disciplines have developed a different understanding of the notion of a model, of the function of models in scientific research and of the purpose of the model. Many different notions are used, e.g. [4,12,18]. There is however not yet a general notion of a model. Our definition of a model [32] summarises the bottom-up approach to models and modelling developed at CAU Kiel.

Models are often language based. Their syntax uses the namespace and the lexicography from the application domain. Semantics is often implicit. The lexicology can be inherited from the application domain and from the discipline. Models do not need the full freedom for interpretation. The interpretation is governed by the purpose of the model within the research scenario, is based on disciplinary concerns (postulates, paradigms, foundations, commonsense, culture, authorities, etc.) and is restricted by disciplinary practices (concepts, conceptions, conventions, thought style and community [5], good practices, methodology, guidelines, etc.). Models combine at least two different kinds of meaning in the namespace: referential meaning establishes an interdependence between elements and the origin ('what'); functional meaning is based on the function of an element in the model ('how'). The pragmatics of a model depends on the community of practice, on the context of the research task and especially on the purpose or function of the model.

2.1 A Model is a Well-Formed, Adequate and Dependable Instrument

A **model** *is a well-formed, adequate and dependable instrument that represents origins.*

Its criteria of well-formedness, adequacy, and dependability must be commonly accepted by its community of practice within some context and correspond to the functions that a model fulfills in utilisation scenarios.

The model should be well-formed according to some well-formedness criterion. As an instrument or more specifically an artifact a model comes with its *background*, e.g. paradigms, assumptions, postulates, language, thought community, etc. The background its often given only in an implicit form.

A well-formed instrument is *adequate* for a collection of origins if it is *analogous* to the origins to be represented according to some analogy criterion, it is more *focused* (e.g. simpler, truncated, more abstract or reduced) than the origins being modelled, and it sufficiently satisfies its *purpose*.

Well-formedness enables an instrument to be *justified* by an empirical corroboration according to its objectives, by rational coherence and conformity explicitly stated through formulas, by falsifiability, and by stability and plasticity.

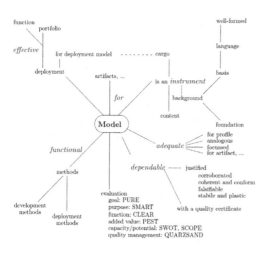

The instrument is *sufficient* by its *quality* characterisation for internal quality, external quality and quality in use or through quality characteristics [31] such as correctness, generality, usefulness, comprehensibility, parsimony, robustness, novelty etc. Sufficiency is typically combined with some assurance evaluation (tolerance, modality, confidence, and restrictions).

A well-formed instrument is called *dependable* if it is sufficient and is justified for some of the justification properties and some of the sufficiency characteristics.

The model has a profile (goal or purpose or function), represents artifacts and is used for some deployment scenario. As an instrument, a model has its own background (e.g. foundation (paradigms, postulates, theories, disciplinary culture, etc.) and basis (concepts, language, assumptions, practice, etc.)). It should be well-defined or well-formed. Adequacy is based on satisfaction of the purpose, analogy to the artifacts it represents and the focus under which the model is used. Dependability is based on a justification for its usage as a model and on a quality certificate. Models can be evaluated by one of the evaluation frameworks. A model is functional if methods for its development and for its deployment are given. A model is effective if it can be deployed according to its portfolio, i.e. according to the tasks assigned to the model. Deployment is often using some deployment model, e.g. for explanation, exploration, construction, description and prescription.

A model can be used for different purposes and various usage scenarios. Therefore, a model is typically also extended by *views* or *viewpoints* that reflect certain parts of the model and that hide details which are not necessary. This reflection is often only provided in a non-systematic or implicit way. Additionally, we need a refinement notion, methods for combination and for evaluation of models.

2.2   Models as a Means in Research and Engineering Networks

A common understanding of the nature of models, of the methods and techniques that are used for model development and model deployment and of systematic approaches to modelling enables also model-based collaboration in networks.

Models are built and modelling is performed in a similar form, with similar background and theories and within similar investigation scenarios despite the variety of models, the variety of purposes, the complexity, the range from micro to macro, and the variety of solutions. Each discipline has been developing also specific solutions to modelling and model deployment. These solutions may also be used for other disciplines, may be combined with their solutions, or may replace their solutions.

Our notion of the model has been validated and verified against the model notions of many disciplines. The validation [33] brought an insight into the specific understanding of adequacy, dependability, functioning, and effectiveness used in each of these disciplines. The validation has also resulted in an understanding of the added value of the model within the discipline, in an evaluation of the model maturity, in detection of features which are missing and should be added to the model or which can be deleted from the model, and in restrictions to model deployment which must be observed.

In Sect. 4 the notion of the model is generalised in order to cope with the requirements for model-based collaboration.

2.3   Local-As-Design for Disciplinary Models

Disciplines have however also their own foundation, their own background, their own culture and their own way of model use. Therefore, it is infeasible to develop a holistic model for everybody in a research team. Models should remain in their local setting and should not be integrated into general global model that is commonly agreed and used. For this reason, we prefer a *local approach*. A model remains within its local environment. It is however enhanced in such a way that it can be used in a collaboration and thus support exchange of ideas and results. This local approach is similar to the global-as-view integration approach used for integration of database systems. The model enhancement needs however a *generalised-global-as-view* approach. The collaboration environment thus supports a peer-to-peer exchange of exchange sub-models.

3 Model-Based Collaboration in Interacting Communities

3.1 Models — The "Intergalactic" Communication Instrument

Collaboration on the basis of models preserves local models which have sub-models for collaboration activities thus providing an explicit association of derived sub-models. Models vary in their abstraction, their foci and scales, their scopes, their aspects and their purposes. They are deployed in different scenarios and are backed by heterogeneous data with different granularity and at different levels of abstraction. A model-based collaboration cannot be based on an exchange of models as they are. Models must be fitted to the partner. We use typically parts or abstractions of a model for exchange. This model transformation is not yet performed in a systematic manner. We might however develop an algebra for such model transformations. In this case we can generate derivatives or exchange sub-models of a model.

If a derivative of a model is used for exchange then the model of the partner can incorporate the derivative. The derivative is typically transformed to the model. It is then integrated into the model that is under revision in such collaboration activities. The derivative is associated to a sub-model of the new model. This association can be the basis for future communication. Modelling itself becomes now teamwork.

Models are simultaneously used in interdisciplinary teams for different interleaved purposes. For instance, a conceptual model of an information system is used for construction and inspiration in an implementation phase, for planning and resource allocation, for verification against the requirements model, for optimisation of the structure, for prognosis of behaviour of the system that is under construction, for explanation and understanding its components, and as the basis for system integration. Each of these functions can be used by different stakeholders at the same time. Typically, only some of model elements are of interest to different team members. These members should be better supported by specific views defined on top of the model. These views should be defined in dependence on the viewpoints that are requested by the partner. If the model is changed then these views must also be changed and the change must be communicated in an appropriate form. Model views are therefore exported to partners.

A change in one model may also result in a change to models of collaborators. The changes should be integrateable into the model. The result of integration by a partner should be communicated within a research team. Model views that are derived from one model are imported into model views of another model. Since models might use different languages the model view that is exported by one model must be transformed before integration into another model.

3.2   Model Suites

Model suites are an extension of model ensembles [22] used for distributed or collaborating databases [25].

A model suite [3, 26] consists

- of set of models $\{M_1, ..., M_n\}$,
- of an association or collaboration schema among the models,
- of controllers that maintain consistency or coherence of the model suite,
- of application schemata for explicit maintenance and evolution of the model suite, and
- of tracers for the establishment of the coherence.

Coherence describes a fixed relationship between the models in a model suite.

The collaboration style of a model suite is based on supporting programs, data access pattern, style of collaboration, and coordination workflows. Collaboration pattern generalize protocols and their specification [16].

Let us assume that a model is defined in a language that uses constructors $\mathfrak{C}$ for the structuring and defining a model M, i.e. $M \in Term(\mathfrak{C})$ for the set of all terms defined in $\mathfrak{C}$. These constructor can be combined with an algebra $\mathfrak{A}$ of expressions defined over $\mathfrak{C}$. Typical operations of the algebra are set operations such as union, difference and intersection, constructing operations such as join, projection, selection, nesting (or integration/combination) and unnesting (or disintegration), and abstraction (or more specifically aggregation) operations. This approach to algebras follows approaches for universal algebras [19].

Each operation can be classified as either an *identification-preserving* or an *identification-loosing* one. Identification-preserving operations are, for instance, difference, intersection, nesting, and unnesting. An expression is identification-preserving if all its sub-expressions have this property.

We may use additional identification auxiliaries, i.e. constructions that define together with the given construction an identification-preserving expression.

A sub-model of M can be either defined as a sub-expression of the expression that defines M or as the application of an expression $E(M) \in \mathfrak{A}(\mathfrak{C})$ with one free variable M. For collaboration networks we choose the second approach and call them exchange sub-models. Sub-models can be identity-preserving or identity-loosing. Given any model or sub-model, an expression $E(M)$ defined on this model can also be considered as a mapping from M to the resulting structure $E(M)$. Furthermore, we can use infomorphisms [9, 28] among models. Two models M_1, M_2 are E_1, E_2-*infomorph* though two transformations E_1, E_2 with $E_1(M_1) = M_2$ and $E_2(M_2) = M_1$ if any object o defined on M_i can be mapped via E_i to objects defined on M_j for $i, j \in \{1, 2\}$, $i \neq j$.

4   The Model Capsule

4.1   The Model Capsule $\equiv$ Model $\bigoplus$ Exchange Sub-models

Models are extended by sub-models that are either

1. *abstractions* of the given model similar to roll-up or aggregation techniques used in database technology [17] or
2. *specialisations* to a more specific model similar to refinement techniques used for abstract state machines [1] or
3. *specific viewpoints* of the given model similar to view schemata [29].

Sub-model specification is based on an algebra for abstraction, refinement and filtering. The algebra is also used for transformation of sub-models. Sub-models to be exported to another model can be transformed before becoming imported by another model.

A **model capsule** consists of a main model and many exchange sub-models. An exchange sub-model is either an export or an import exchange-sub-model. If it is an import sub-model then it must be identity-preserving[1].

Exchange sub-models are used as mediator in research teams and provide all details that are necessary for collaboration (completeness) but only those (minimality). Exchange sub-models are either derived from the main model in dependence on the viewpoint, on foci and scales, on scope, on aspects and on purposes of partners or are sub-models provided by partners and transformed according to the main model. A team member thus can integrate an exchange sub-model in his/her main model, can propagate changes made by him/herself to other partners and can change the main model according to changes by partners. This model capsule is the main communication vehicle for collaboration. The propagation and transformation from and to partners can be based on contracts or protocols.

4.2   Collaboration Model Capsules

Model suites can be associated to each other based on exchange sub-models.

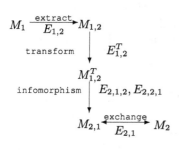

Given model suites $\mathcal{M}_i = (M_i, M_{i,1}, \ldots M_{i,n_i})$ with n_i exchange sub-models. A model suite $\mathcal{M}_1$ is bound to a model suite $\mathcal{M}_2$ via export/import sub-models $M_{1,2}$ and/or import sub-models $M_{2,1}$ if there exist expressions $E_{1,2}$, $E_{1,2}^T$, $E_{2,1,2}$, $E_{2,2,1}$, $E_{2,1}$ such that $E_{1,2}$ extracts the sub-model $M_{1,2}$ from M_1, the transformation expression transforms this sub-model to a model $M_{1,2}^T$ that is infomorph to the import sub-model $M_{2,1}$ of $\mathcal{M}_2$, i.e. formally ·

---

[1] This restriction can be weakened if additional identification auxiliaries are used.

$E_{1,2}(M_1) = M_{1,2},$
· $E_{2,1}(M_2) = M_{2,1},$
· $E_{1,2}^T(M_{1,2}) = M_{1,2}^T,$ and
· $M_{1,2}^T$ and $M_{2,1}$ are $E_{2,1,2}, E_{2,2,1}$-infomorph.

We notice without proof that the infomorphism can be integrated into the transformation expressions for some special cases.

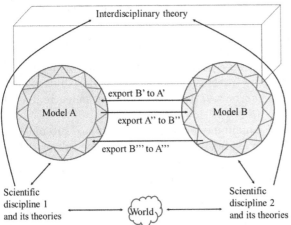

Expressions we use for model association may be, for instance, aggregation or abstraction expressions, viewpoint expressions, specialisation expressions, or also combination expressions. Therefore, a sub-model of a first model that is used for association with a second model may be more abstract, or may be oriented on specific elements of the first model, or may extend the first model. Abstraction allows to form a kind of generalisation, i.e. a vertical hierarchy. The model capsule is bound vertically. Specific or extended models are typically defined on the same level of abstraction. The model capsule is then bound horizontally.

This approach is sufficiently general for model-based communication and reasoning in interacting research and engineering communities. Each branch of engineering or science uses its specific model suite. In order to collaborate, an interdisciplinary theory is formed. The interdisciplinary theory corresponds to the association in the real world. For instance, model capsules are based on models A and B that use corresponding scientific disciplines and corresponding theories as a part of their background. The models have three derived exchange sub-models that are exported to the other capsule and that are integrated into the model in such a way that the imported sub-model can be reflected by the model of the capsule. The two models and the two scientific disciplines are the kernel for an interdisciplinary theory.

5   Realisation and Implementation of the Approach

Model suites have already been investigated for UML-based software engineering in [26] on the basis of [30]. M. Skusa investigated the association among modelling languages based on language mappings. Each of the diagram types got its own profile. These profiles have been used for automatic derivation of associations among UML diagrams. The direction of enforcement follows in this case waterfall development strategies, i.e. requirements diagrams cannot be changed by conceptual diagram changes. He also developed controllers that maintain consistency of diagrams within a model suite. These controllers have been written as rules based on Abstract State Machines [1]. Since ASM rules run in parallel all controller run in parallel.

The Extract-Transform-Load paradigm can be enhanced by derivation of functions that provide the basic database system CRUD functionality [34]. Therefore, exchange sub-models support database processing similar to classical technology.

Traditional object-relational approaches only support singleton table views. To overcome this limitation we define a complex view as a collection of views that are associated through integrity constraints - mainly (pairwise) (generalised) inclusion constraints. The view classes are computed in the first step from the basic database using the view expression and then mapped to a database based on the association schema. They thus form a local database on their own.

The concept of view towers [15] has already been used for the generation of interfaces. Views of level i are schemata on their own and are incrementally constructed of the base database schema (level 0) and of views of level less than i. It has been shown that SQL and database technology nicely support such complex views [13]. The construction of view towers can be enhanced by a characterisation whether the view is updatable. A higher level view is strongly updateable if the algebraic expression that defines this view does not destroy updateability and each of its components of lower level is updateable. Views can be enhanced by auxiliary views that provide an enhanced updateability based on a combined view of the original one and the auxiliary ones that is itself updateable.

5.1 Realisation 1: Applicability of the Approach in Research Communities

Our approach supports collaboration for more complex applications discussed in Subsect. 1.1. We define explicit transformation expressions. The notion of infomorphism becomes then however far more complex. Both research collaborations in the Clusters of Excellence at CAU Kiel are using ad-hoc model associations. In [6] model suites and views have been used for automatic recharge of archives. The development and maintenance of integrated, reusable and coherent archives for all data capturing project is a mandatory requirement issued by the German Research Council to integrated projects such as Clusters of Excellence. In [7], a general data store has been realised for all archeology and pre-historic data. The local projects have their import views to and export views from the general data store. The global data store consists of one component. This component contains all data from all projects in the Graduate School "Human Development of Landscapes" and a pair of an import and an export view for each of the projects. Project collaboration is based on collaboration export views for each collaborating community. The projects themselves have their own database schemata that correspond to the import view through an extract-transform-load feature.

Both database support projects are the basis and the background for the model capsule approach developed in this paper.

5.2   Realisation 2: Collaboration Model Capsules in Software Engineering

Let us now exemplify the concept for classical software engineering with an example adapted from [27].

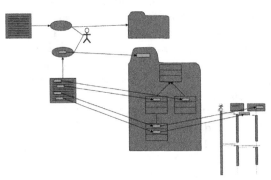

Given a use_case diagram, a class diagram, a package diagram, and an interaction diagram. These four diagrams can be associated by exchange sub-model for a use_case-package association in the upper part and package-interaction, package-class, and class-use_case associations in the lower part. Controllers maintain the coherence of the different viewpoints. In the lower part, we consider the package diagram to be the leading diagram for the class and interaction diagrams and the class diagram as a leading diagram for the use case diagram. The class diagram has an export sub-model to the use_case diagram that has an identity-preserving sub-model as an import sub-model. Controllers may use a restrict, eager or lazy approach, i.e. a change in the class-diagram is allowed

- only if this can be directly reflected in the use_case sub-model (restrict) or if this changed be directly (eager) or at a later stage (lazy) propagated to the importing sub-model of the use_case model and
- the change modifies the export sub-model in the class diagram. The application schemata are derived from controllers based on templates or pattern similar to integrity maintenance for referential inclusion constraints in databases. Tracers are then small demons that observe whether a model changes its export and import models.

6   Conclusion

This paper proposes an approach to models and model-based reasoning for interacting research and engineering communities. Models are an *"intergalactic" communication and reasoning instrument* and a *crosscutting concern* in such networks. Model-based communication and reasoning is based on the concept of the model capsule that provides a flexible and powerful mechanism for model-based reasoning and collaboration.

They provide a flexible system to manage locally and to exchange globally models for collaboration in teams. Models thus become a crosscutting concern to reflect competence for an interdisciplinary collaboration and for interactive research on complex society issues that cannot be solved within a singleton discipline.

The role and potential of models in networked research communities has not yet been systematically investigated, explored and generalised. This paper tackles the collaboration challenge based on model-based data exchanging collaboration. Model-based collaboration is only one kind of collaboration beside the data-based, concept-based, workpiece-based, process-oriented etc. collaborations. It seems however that models are a central instrument for any qualified and dependable collaboration.

As the next step, we aim at a general model description language ModelML that allows to collect models in networks in a form similar to an online interactive encyclopedia or model web. This model web supports systematic elicitation and exploration of modelling experience in research networks.

References

1. Börger, E., Stärk, R.: Abstract State Machines - A Method for High-Level System Design and Analysis. Springer, Berlin (2003)
2. Coleman, A.: Scientific models as works. Cataloging Classif. Q. Spec. Issue: Works Entities Inform. Retrieval **33**, 3–4 (2006)
3. Dahanayake, A., Thalheim, B.: Co-evolution of (information) system models. In: Bider, I., Halpin, T., Krogstie, J., Nurcan, S., Proper, E., Schmidt, R., Ukor, R. (eds.) BPMDS 2010 and EMMSAD 2010. LNBIP, vol. 50, pp. 314–326. Springer, Heidelberg (2010)
4. Dirks, U., Knobloch, E. (eds.): Modelle. Peter Lang (2010)
5. Fleck, L.: Denkstile und Tatsachen. In: Werner, S., Zittel, C. (eds.) Surkamp (2011)
6. Fleischer, D., Jannaschk, K.: A path to filled archives. Nat. Geosci. **4**, 575–576 (2011)
7. Förster, F., Thalheim, B.: An effectual approach for a data and information management for humanists. Qual. Quant. Methods Libr. (QQML) **2**, 121–128 (2012)
8. Gillett, N.P., Zwiers, F.W., Weaver, A.J., Hegerl, G.C., Allen, M.R., Stott, P.A.: Detecting anthropogenic influence with a multi-model ensemble. Geophys. Res. Lett. **29**(20), 31–34 (2002)
9. Goguen, J.: Three perspectives on information integration. In: Kalfoglou, Y., et al. (eds.): Semantic Interoperability and Integration. Dagstuhl Seminar Proceedings 04391 (2005)
10. Gray, J., eScience: A transformed scientific method. Technical report, Talk given Jan 11, Edited by Hey, T., Tansley, S., Tolle, K. (2007). http://research.microsoft. com/en-us/um/people/gray/talks/NRC-CSTB_eScience.ppt, Microsoft Research Publications
11. Guerra, E., de Lara, J., Kolovos, D.S., Paige, R.F.: Inter-modelling: from theory to practice. In: Petriu, D.C., Rouquette, N., Haugen, Ø. (eds.) MODELS 2010, Part I. LNCS, vol. 6394, pp. 376–391. Springer, Heidelberg (2010)
12. Halloun, I.A.: Modeling Theory in Science Education. Springer, Berlin (2006)
13. Holst, F.: Konzeptuelle Sichten. Master's thesis, CAU Kiel, Institut für Informatik, Kiel, Juli 2015

14. Hunter, P.J., Li, W.W., McCulloch, A.D., Noble, D.: Multiscale modeling: phys-iome project standards, tools, and databases. IEEE Comput. **39**(11), 48–54 (2006)
15. Jaakkola, H., Thalheim, B.: Multicultural adaptive systems. In: Information Mod-elling and Knowledge Bases XXVI. Frontiers in Artificial Intelligence and Appli-cations, vol. 272, pp. 172–191. IOS Press (2014)
16. König, H.: Protocol Engineering: Prinzip. Beschreibung und Entwicklung von Kom-munikationsprotokollen. Teubner, Stuttgart (2003)
17. Lenz, H.-J., Thalheim, B.: A formal framework of aggregation for the OLAP-OLTP model. J. UCS **15**(1), 273–303 (2009)
18. Magnani, L., Carnielli, W., Pizzi, C. (eds.): Model-Based Reasoning in Science, Technology: Abduction, Logic, and Computational Discovery. Springer (2010)
19. Malzew, A.I.: Algebraic Systems. Nauka, Moscow (1970)
20. Pardillo, J.: A systematic review on the definition of UML profiles. In: Petriu, D.C., Rouquette, N., Haugen, Ø. (eds.) MODELS 2010, Part I. LNCS, vol. 6394, pp. 407–422. Springer, Heidelberg (2010)
21. Pottmann, M., Unbehauen, H., Seborg, D.E.: Application of a general multi-model approach for identification of highly nonlinear processes - a case study. Int. J. Control **57**(1), 97–120 (1993)
22. Prochnow, S., von Hanxleden, R.: Statechart development beyond WYSIWYG. In: Engels, G., Opdyke, B., Schmidt, D.C., Weil, F. (eds.) MODELS 2007. LNCS, vol. 4735, pp. 635–649. Springer, Heidelberg (2007)
23. Rumpe, B.: Modellierung mit UML. Springer (2012)
24. Salay, R., Mylopoulos, J., Easterbrook, S.: Using macromodels to manage collec-tions of related models. In: van Eck, P., Gordijn, J., Wieringa, R. (eds.) CAiSE 2009. LNCS, vol. 5565, pp. 141–155. Springer, Heidelberg (2009)
25. Schewe, K.-D., Thalheim, B.: Development of collaboration frameworks for web information systems. In: IJCAI 2007 (20th International Joint Conference on Arti-ficial Intelligence, Section EMC 2007 (Evolutionary models of collaboration), pp. 27–32, Hyderabad (2007)
26. Skusa, M.: Semantische Kohärenz in der Softwareentwicklung. Ph.D. thesis, CAU Kiel (2011)
27. Skusa, M., Thalheim, B.: Kohärente Multi-Modell-Entwicklung. In: [31], pp. 431–454
28. Song, D., Bruza, P.: Discovering information flow using a high dimensional concep-tual space. In: Research and Development in Information Retrieval, pp. 327–333 (2001)
29. Thalheim, B.: Entity-Relationship Modeling - Foundations of Database Technol-ogy. Springer, Berlin (2000)
30. Thalheim, B.: The conceptual framework to multi-layered database modelling. In: Proceedings of the EJC, pp. 118–138, Maribor, Slovenia (2009)
31. Thalheim, B.: Towards a theory of conceptual modelling. J. Univers. Comput. Sci., 16(20), 3102–3137 (2010). http://www.jucs.org/jucs_16_20/towards_a_theory_of
32. Thalheim, B.: The conceptual model ≡ an adequate and dependable artifact enhanced by concepts. In: Information Modelling and Knowledge Bases XXV. Frontiers in Artificial Intelligence and Applications, vol. 260, pp. 241–254. IOS Press (2014)
33. Thalheim, B., Nissen, I. (eds.): Wissenschaft und Kunst der Modellierung. De Gruyter, Ontos Verlag, Berlin (2015)
34. Tropmann, M.: Funktionale Integration heterogener Datenbanken. Master's thesis, CAU Kiel, Institut für Informatik, Kiel, Januar 2008

ADBIS Doctoral Consortium

Usage of Aspect-Oriented Programming in Adaptive Application Structure

Jiří Šebek[✉] and Karel Richta

Department of Computer Science and Engineering, Faculty of Electrical Engineering,
Czech Technical University in Prague, Karlovo nám. 13, 121 35 Prague 2, Czech Republic
{sebekji1,richta}@fel.cvut.cz

Abstract. Adaptive Application Structure (AAS) is one that changes its structure based on the current context. It brings benefits to end users, cause everyone has its own needs. Also the applications is created in some way by developers, but each user is using the application in different way. In AAS-approach development difficulties appears caused by extended development and maintenance efforts. to implement. This paper considers the Aspect-Oriented Programming (AOP)-based approach for the AAS design.

Keywords: Adaptive application structure · Aspect-driven design · Entity inspection based approach · Run-time aspect model · Reduced maintenance and development efforts

1 Introduction

Software applications provide UI for the interaction with users. These days User Interfaces (UI's) are very important and their primary goal is to offer the best possible comfort service. There arises a problem with increasing number of different users with special requirements and abilities. The number of different electronic devices and the number of different environments with different types of users grows. It is necessary to create specific UIs that can adapt to the particular user or more generally a context. The context characterizes the situation where the application is used. It may include the current environment, device properties, by which user controls the application and the capabilities and preferences of the user. [7] defines context as:

Context is any information that can be used to characterize the situation of an entity. An entity is a person, place, or object that is considered relevant to the interaction between a user and an application, including the user and applications themselves.

Also there are another definitions. The definition that Dey provided [7] is the most general one. The first definition that were made by Schmidt [7] is defined as:

This work has been partially supported by the Grant Agency of CTU No. SGS15/210/OHK3/3T/
13 and partially also by the AVAST Foundation

© Springer International Publishing Switzerland 2016
M. Ivanović et al. (Eds.): ADBIS 2016, CCIS 637, pp. 217–222, 2016.
DOI: 10.1007/978-3-319-44066-8_22

Context awareness as knowledge about the user's and IT device's state, including surroundings, situation, and to a less extent, location.

The software structure is equally important as UI. Not only that user wants to see the concrete information for him but also he wants to find it fast. In general it is impossible to predict what user wants to do or gets from application and render that information in the first screen of application. One user can use the application as whole unit. So all functions are used by user. The other user wants to use only some of the application functions. Here the application structure should be more *user-friendly* for functions that user wants.

The applications that are using context information (e.g. for adaptation) is called Context-aware. The AAS has similar problems as Adaptive User Interface (AUI). Both uses context to adapt structure or in AUI case UI part of application. For a context dependent application it is necessary to expend a lot of resources to the development and maintenance of the AAS or AUI descriptions. The main problem is the complexity, content volume and the number of interactive elements. An example of this problem is the construction of a multiple platform application. We often have to create several separate applications or subsystems, because there are different technologies or different interaction with a particular device. This results in increased costs, low flexibility to contextual situations and complex adaptation to changes of the underlying application. These changes must be done separately for each platform.

The motivation for this paper is to define AAS problem, show its strong points. Also the paper will come up with difficulties in development in this approach.

2 Background

As we defined the term context in the introduction, it can be divided into four parts [1]. Classification is divided into two categories according to the type of changes to *static* and *dynamic*. If the parameter is changed during the session, it is assigned a *dynamic* parameter, otherwise as *static*. The next level of classification is divided into *functional* and *presentation*. The *functional* category contains parameters that change functionality of backend. The second category contains parameters that change the frontend. Table 1 shows an example.

Table 1. Placement of context parameters into categories

	Static	Dynamic
Functional	Role, identity, platform, preferences	Connection to the internet, time, location
Presentation	Platform, preferences	Lighting, activity of user, orientation of device

2.1 Aspect-Oriented Programming (AOP)

AOP [6] is a programming paradigm that deals with separation of concerns. Concern is any functionality or behavior of the program. The result of this is separation of the program into functional section that is not overlapping in the best case. The OOP separates the concerns into classes, based on the implementation. AOP is more capable.

It addresses also separation of cross-cutting concerns. The cross-cutting concerns are concerns that cannot be encapsulated easily and they appear in multiple places in program. AOP deals with these sorts of concerns and separates them from the program into aspect modules.

3   Related Works

In this chapter there are described adaptation software development approaches important for this paper. Here there are described options of code generation - Meta-programming (MP), Model-driven development (MDD).

3.1   Code Generation Approaches

Meta-programming (MP) brings the possibility to a program to observe, generate, combine, manipulate and transform its elements [2]. MP code generators usually combine code fragments together to construct a functionality or to adapt to unknown components. Code analyzers use MP to observe the structure of a program to compute a value.

Model-Driven Development (MDD) [3] is a paradigm that uses models in software development. Models can be represented in visual or textual format (e.g. UML or XML). In MDD there are different models for different level of abstraction. Model transformations are used to transform a model to another model or in final step into program code. Model that described another model is called meta-model.

[4], similar to AOP addresses cross-cutting concerns. It combines domain specific descriptions, models, code, etc. through a transformation involving templates. Its output resolves at compile time and compare to AOP does not provide well-described formalization and join mechanism.

4   Structure Adaptation

In the most cases the application structure is static. In order to create adaptation for the structure it is natural to save it in some meta-model. For this approach we can use AOP. It will reduce our effort to develop such application. When we have set some application structure and save it in meta-model (in the most cases it is some type of cache). Now how to display it in the way user wants. Information grouping is term which specifies the process of information gathering into groups. The simple applications can have structure that can be represented by simple list. The more complex ones can be represented by:

- Menu
- Tab folders
- Collapsible blocks
- Different dialogs with navigation
- Wizard based UI
- Tree view

From the list above we can see that this list is primarily targeting desktop applications.

In Android mobile application there is structure called Fragments instead of Tab folders. Collapsible blocks are very unacceptable in mobile platform. There is small space for layout, UI elements and it should not move to disturb or confuse the end-user. Different dialogs with navigation has same problem on mobile platform also. The contexts are different. In the web applications we can get some information from browser cache. Mobile devices can give us more information (e.g. from sensors, devices hardware). The most used tool to represent structure is menu. In the list below there is classification based on end-device:

– Desktop (Web) application
– Mobile application

The meta-model of application structure can be represented by graph. A graph is made by two sets called *Vertices* and *Edges*. It may look like tree but tree is undirected graph and as we can see in Fig. 1 on the left side it is directed graph. We can see the similarity between menu structure and tree structure. In Fig. 1 on the right side we can see the menu we will not considering as structure of menu. The main reason is fact, why should developer make a loop in menu like this. It is redundant link and user will just be confused why there is same item placed in two spots. We have to take a note the structure we are presenting in this paper is navigation structure. If there is a link inside the node in menu (program screen) it is alright but do not need to visualize this information in navigation structure.

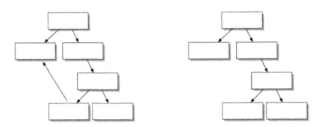

Fig. 1. Graph representations of application structure

Vertices are represented by nodes. Nodes is divided into: *root, inner* and *leaf. Root* is only one per menu. *Inner* node can represent just group information or clickable link to page. *Leafs* are always links.

4.1 Mobile Platform - Android

In the chapter above it was mentioned types of nodes. We can divide this into more detailed by platform classification:

– Desktop (Web) application
  • Inner node can represent just group information or clickable link to page.
– Mobile application
  • Inner node can represent just group information not clickable link to page.

Another limit comes up on Android platform. The depth of menus that can be created is only 3 (if we count root as well).

In this paragraph we can define two instances of this structure adaptation:

– User that uses only specific branch of nodes
– User that uses all nodes

User from the first category needs to reach the used nodes fast as possible. So if it is possible we need to move them up in our structure. Also there is subproblem. If there is a lots of leaf nodes in one place. This long list is hard to list and time consuming specially on mobile devies. The question is if solution should divide this leaf nodes from one parent into two parts and create another parent node for the second part. The problem will come up in naming the second parent. Also how should user recognize in which of these two parent nodes will he find his leaf node. User from the second category needs to reach all nodes in the same average time. So if it is possible we need to adjust structure to be as ballanced as possible. In Fig. 2 there are two ideal structures. The gray filled nodes are ones that user is using. On the left side of Fig. 2 there is structure for user that is using only one node. As we can see this node is as much on the top of the structure. The rest of structure can remain same. On the right side of Fig. 2 there is ideal structure for user that is using all nodes. This show just basic concept. If there will be more nodes it will represents list and it is not ideal. For more nodes it is required to let some part of original structure there and create more complex structure that will be balanced.

Fig. 2. Ideal structure for specific type of users

The same priority as solving the problem itself has also how should work the algorithm in lifetime of application. If the structure will change every time user will start application it will confuse him and user will refuse to use it. It is needed to set number of application starts and then finds out how adjust application structure. Or it is needed to set period in which will framework gather information. The first choice seems like to be the better one, cause we can be sure that user used application. In the second choice we do not know if user use it or how much. As a result algorithm will have two phases - *Learning phase* and *Production phase*. This princip of two phases is used in distributed algorithm specially in *Leader election*. In second phase there will be also short period of users learning phase. User will have to adapt for the better structure cause he will be adapted to the old structure.

Learning phase has some difficulties too. We do not know how to find out if users behavior was exception and when it is users ordinary behavior pattern. The boundary is not known for now.

5 Conclusion and Future Work

In this paper, we have presented ways how to handle AAS and possible extension of AOP-based framework from [5]. There are lots of classifications of this problem. We demontrate the main ideas of this methodology with AOP-based approach. The future work will have two important steps. The first one will be to implement extension for desktop applications and then for mobile applications. The second important step will be tested with real users. The expectation is that applications with AAS will be more *user-friendly* and users will have better feeling while using it.

References

1. Černý, T., Čemus, K., Donahoo, M.J., Song, E.: Aspect-driven, Data-reflective and Context-aware User Interfaces Design. In: Applied Computing Review, vol. 13, Issue 4, pp. 53–65. ACM, New York (2013). ISSN 559-6915. http://www.sigapp.org/acr/Issues/V13.4/ACR-13-4-2013.pdf

2. Lopez, M., Reis, G.D.: Meta-programming with welltyped code analysis. In: Proceedings of the 30th Annual ACM Symposium on Applied Computing, ser. SAC 2015, pp. 2119–2121. ACM, New York (2015). http://doi.acm.org/10.1145/2695664.2696012

3. Van den Bergh, J., Meixner, G., Breiner, K., Pleuss, A., Sauer, S., Hussmann, H.: Model-driven development of advanced user interfaces. In: CHI 2010 Extended Abstracts on Human Factors in Computing Systems, ser. CHI EA 2010, pp. 4429–4432. ACM, New York (2010). http://doi.acm.org/10.1145/1753846.1754166

4. Czarnecki, K.: Overview of generative software development. In: Banâtre, J.-P., Fradet, P., Giavitto, J.-L., Michel, O. (eds.) UPP 2004. LNCS, vol. 3566, pp. 326–341. Springer, Heidelberg (2005). http://dx.doi.org/10.1007/11527800n25g

5. Šebek, J., Richta, K.: Aspect-oriented user interface design for android applications. In: DATESO 2015, pp. 121–130. Matfyspress, Prague (2015). ISSN 1613-0073, ISBN 9788073782856

6. Kiczales, G., Lamping, J., Mendhekar, A., Maeda, C., Lopes, C., Loingtier, J.-M., Irwin, J.: Aspect-oriented programming. In: Akşit, M., et al. (eds.) ECOOP 1997. LNCS, vol. 1241, pp. 220–242. Springer, Heidelberg (1997). http://dx.doi.org/10.1007/BFb0053381

7. Abowd, G.D., Dey, A.K., Brown, P., Davies, N., Smith, M., Steggles, P.: Towards a better understanding of context and context-awareness. In: Gellersen, H.-W. (ed.) HUC 1999. LNCS, vol. 1707, pp. 304–307. Springer, Heidelberg (1999)

Short-term User Behaviour Changes Modelling

Ondrej Kassak[(✉)], Michal Kompan, and Maria Bielikova

Faculty of Informatics and Information Technologies,
Slovak University of Technology in Bratislava,
Ilkovicova 2, Bratislava, Slovakia
{ondrej.kassak,michal.kompan,
maria.bielikova}@stuba.sk

Abstract. As the Web becomes more and more dynamic, it is interesting to explore the short-term modelling of its user behaviour. Nowadays, it is important to have an information about user's preferences and needs online. It allows us, in addition to other advantages, also to predict user's future actions. In this paper we describe the doctoral research focused on the modelling of the short-term changes in user's behaviour. We explore the task of user session exit intent prediction. Our approach employs generally available data sources on user behaviour on the Web, so it is domain independent.

Keywords: User modelling · Session exit intent prediction · Web mining · Short–term user behaviour

1 Introduction and Related Work

Users interacting with a web-site produce high amount of footprints. These are typically logs describing simple actions, e.g., page visits or hyperlink clicks. Speaking generally, such data bring us only low information value. However, logs gather much more - individual actions made by a user with some intent, in a certain sequence, in specific time and duration and much more. Some pages are popular and a user returns often, some he/she visits only once, thanks to influence of actual circumstances.

User modelling is a well-known research area, which helps us to understand the previous behaviour of specific user as well as the whole communities. Thanks to the modelling we are able to personalize the web-sites or to improve the user experience. Moreover, the site structure or its informational architecture can be improved.

Traditionally, user modelling focuses on user's long-term preferences and his/her stable interests. It allows us to personalize web-site content – typically by recommending interesting items to user [3]. In this context, the modelling of users' preferences is often extended to group modelling [4], which allows to adjust sites for purposes of multiple users at once.

Long-term preferences modelling is however insufficient for tasks where the dynamic reaction to slight changes in user behaviour is needed. As the Web become more dynamic and its content changes rapidly (e.g., social networks, news), it is important to model also the user's actual intent and his/her actual preferences. To put it simply, the short-term behaviour is important in the Web context. For this purpose,

© Springer International Publishing Switzerland 2016
M. Ivanović et al. (Eds.): ADBIS 2016, CCIS 637, pp. 223–228, 2016.
DOI: 10.1007/978-3-319-44066-8_23

there were proposed approaches modelling user interests on multiple time intervals. Typically, there exist two level approaches modelling preferences on long- and short-term level [8]. Some authors proposed multiple level models [11], which provide additional details. All of these approaches share one feature – they separately capture long- as well as short-term user's preferences. This allows to observe at the same time stable and also short term actual interests and preferences.

This information helps to analyse and understand user's past behaviour. On the other side, it can be used to predict the future behaviour. It also can be used for prediction of the next user's action within the page or the intent to leave the web-site.

The task of user exit or loss was previously researched mostly on the long-term level in the mean of the customer loss [9] or student course fail [10]. In some situations, it is however beneficial to predict the moment when the user leaves his/her actual session. This task was previously researched sporadically – as the exit from tutorials reading [7], recognition of shopping cart abandonment [6] or leaving the search session without obtaining a result. In our previous work we researched the task of the session exit intent prediction [2]. It is very useful because an early indication that the user is about to leave the session, can help the web-site provider to motivate him to stay longer and perhaps, to make some profit.

In the case of the session exit intent prediction, we identified a lack of existing user modelling approaches. Capturing the changes in user's behaviour on the level of actions is a difficult task, which requires understanding of such a complex process. It may seem that such a task requires a high amount of data describing various aspects of user's behaviour, e.g., his/her actual intent, context or previous knowledge. This information is however in real-world applications available only rarely. For this reason, in our research we compensate it by observation of user's behaviour in actual session and in previous sessions for various time intervals (e.g., last hour, day, week).

2 Mining User Behaviour

In our research we focus mainly on commonly available data sources (e.g., simple server logs). In this way, our methods can be applied to a wide scale of existing web-sites and applications. For the purpose of short-term user behaviour modelling from the limited data sources, it is however need to insure the data quality and mine important the information based on the users' session data.

2.1 Web Site Mining

The web-site mining is typically divided into three basic tasks – Web structure, content and usage mining [5]. The first task is used for acquiring the site structure, namely the identification of web-pages and their mutual hyperlink connections. The content mining is focused to individual pages - their textual or multimedia content. Data mined from these sources explain primarily users' preferences and their favourite concepts. In the case of data content, it is important to consider the language dependence and domain variability of mined web-sites. In addition, as the page content evolve in time, there is needed to consider also its temporal aspect (mine the page content that user saw in time of his/her visit).

For purposes of the user's behaviour modelling, the usage mining is the most important source of information. It is typically captured in the form of user's visits of individual web-pages. The actions individually do not offer additional information value –only the target page and time of user visit can be deduced. For this reason, the actions are typically joined into bigger logical units called sessions. Several approaches how to join actions into sessions are used, often the temporal heuristic characteristic brings sufficient results due to its simplicity and precision (ensured mainly by long-tail distribution of session length in time) [1].

In contrast to single user's visit, the user's session adds high amount of information on top of it, which would be hidden in actions considered individually. In the case of session, the actions together reveal e.g., an actual user intent (topic of the actions shared). The user sessions observed for a certain time show in some extent also regularity and uniformity of the behaviour. Again, the user regularly visits the pages referring to his/her long-term preferences, he/she spends similar time in sessions, visits categories in similar order (in news site) etc. For this reason, it is suitable to observe user's behaviour on the level of sessions – from the short- as well as long- term point of view.

2.2 Data Pre-processing

To be able to model user's behaviour, the pre-process of data obtained from server logs have to be performed. In this way, several noisy factors are reduced. As the logs reflect all actions over the Web system we need to filter out the outliers at first. This means to remove the users who behave highly different from the other users. In the case of Web applications these are most typically various crawlers and other automatized bots that perform visits in the high amount and produce unnaturally long sessions. The user behaviour is often compared with the behaviour of others in the user modelling, thus not omitting of outliers would lower the model quality also for other users.

The second step of data pre-processing is typically an attributes normalisation. In modelling are used various data, which are measured often in different scale (e.g., session time in seconds, TF-IDF of visited topic in interval $<0;1>$), the differences in attributes variance can highly differ. Because of further usage of user model, when the machine learning approaches are often used, the normalisation brings positive aspects.

3   Modelling of User Short-term Behaviour

In our research we focus on the modelling of changes in user's short-term behaviour. This task differs from the well-researched modelling of user preferences, because it captures slight different aspects of user's behaviour. We attribute features as how much time a user spends on the page, which action of the sequence it was or which actions were previous. In other words, the content-based aspects are almost ignored.

We proposed a domain independent user model used for capturing the short-term behaviour. Its principle is based on the observation of actual user session, his/her previous sessions for various time intervals and also sessions of other users. In this way it is possible to compare the actual session with the previous from long- as well as short-term point of view. Thanks to this comparison, we are able to express the similarity of actual user's behaviour in certain time and his/her behaviour in the last week or the behaviour of other users in the last month. In our approach used following features:

- number of actions (categories) visited in the session
- time spent
- probability of leaving the session after visiting n most recently visited pages (categories).

For the comparison of actual session with previous sessions in various time intervals we capture:

- number of actions (time spent) made in the actual session divided by average number of actions made in previous sessions
- overall probability of leaving the session after visiting actual page (category) based on previous sessions
- binary flags, describing if the number of actions (time spent) made in the actual session is longer than average number of actions made in previous sessions.

In this way we effectively express how much the user's behaviour in a certain time reminds the previous (already ended) sessions and thus how probable is that the user leaves his/her actual session soon. Moreover, this principle helps with the authentication of users based on the session actions by a comparison of anonymous session with the sessions of known users (e.g., if the user forgot to log in or his/her cookie expired).

Our user modelling method is based strictly on the behavioural data. This ensures the independence from domain and also from the language. The result is thus a widely usable user model. It is however unable to capture any information about content or structure of modelled web-site. Including these features (structure or some abstraction of the site content) can bring further extending of our method a help to better describe how and why the user's behaviour changed. As a result, actually we capture how much time the user spent on the web-page, but we do not know why. In the case we know the page content, we would be able to answer this question more properly (e.g., length of the article, which user reads).

Considering the site content does not necessary mean a loss of the domain independence. The content can be described via indirect characteristics, e.g., the text length, number of paragraphs or number of images.

4   Application of Proposed User Model

We experimentally evaluated proposed user model (describing changes in user behaviour in short-term level) on the task of the user session exit intent prediction. In other words, we predicted if a user will exit the web-site and leave the session in his/her next step or if he/she will continue browsing.

For this task we proposed a prediction method using linear regression classifier which learned importance of model features characteristics. These weights (importance), are in the online time learned by stochastic gradient descent algorithm. In this way the prediction method dynamically reacts to changes in the user's behaviour and preferences [2].

We considered the prediction task in multiple variants. At first we evaluated basic scenario of prediction exit in the next action, in addition we explored the prediction within 1 up to 3 actions in advance.

We experimented with two domains – e-learning and news. In e-learning, we used dataset of 880 users, 460 k actions and 31 k sessions. We reached precision 59.2 for the task of prediction session exit in the next action and even precision 81.9 for the prediction of three actions in advance. We found out that it is difficult to predict exact moment of the exit, but we are able to well predict the group of candidate actions, from which the user will exit probably.

The results for the news domain follow similar pattern as it was in e-learning domain. They however slightly differ due to the different characteristics of the sessions in news domain. In the dataset used for experiments (3 k users, 100 k actions, 50 k sessions) extremely short sessions of 1–2 actions prevail. In such short sessions it is almost impossible to predict any changes in behaviour. We reached the precision 58.0, for the task of exit in next action and precision 74.5 for exit in two actions in advance. The prediction of three actions is advance was not evaluated due to the length of sessions.

Proposed user model showed as a suitable source of information for defined prediction task. The prediction of the user's close exit intent seems to be a promising result for many business applications. Such information brings opportunity to react from the web-site side and there is a higher chance monetize such visit (not only from the business, but also from the user point of view).

In our user model, an actual session is, in addition to his/her previous own sessions, compared also with sessions of other users. In this way it is possible for example to react to the global trends or the domain seasonality. As the users behave heterogeneously, it can lower the quality of modelled information and further the prediction also. This disadvantage could be reduced by the by selection of subset of similar users instead of actually used all users used for sessions comparison. To be able to use similar users, we need to define a similarity measure specified for this task (method able to calculate the user similarity in online time).

5 Future Direction of Research

Modelling of changes in the user short-term behaviour represents a novel task of capturing the user's behaviour on the Web. It is a direct answer to increasing dynamics of web-sites. Nowadays it is important to react to behaviour changes in online time and to use such knowledge immediately to increase the user's browsing experience. For this reason, we explore the task of short-term modelling and its application for the user session exit intent prediction. We found that even from simple server logs (including minimum of information about user previous actions) it is possible to successfully predict the user exit within few actions in advance.

Proposed method is based strictly on the web usage data. This results to its domain and also language independence. Consideration of indirect content and structural characteristic of the web-site, should however help us to mine multiple information from the available input data. We believe, that it will increase the quality of the model and also prediction. Our primary aim is to keep model independencies. Moreover, we believe that the consideration of the similarity measure between the users will bring the improvement of the modelled data quality (by comparing the actual user session with the other user sessions).

Acknowledgement. This work is partially supported the grants APVV-15-0508, VG 1/0646/15 and it is the partial result of the Research and Development Operational Programme for the project No. ITMS 26240120039 co-funded by the ERDF and STU Grant scheme for Support of Young Researchers.

References

1. Herder, E.: An Analysis of User Behavior on the Web - Understanding the Web and its Users. VDM Verlag, Saarbrücken (2007)
2. Kassak, O., Kompan, M., Bielikova, M.: Student behavior in a web-based educational system: exit intent prediction. Eng. Appl. Artif. Intell. J. **51**, 136–149 (2016). Issue Mining the Humanities: Technologies and Applications, Elsevier
3. Kassak, O., Kompan, M., Bielikova, M.: Personalized hybrid recommendation for group of users: top-n multimedia recommender. Inform. Process. Manage. J. **52**(3), 459–477 (2016). Elsevier
4. Kompan, M., Bielikova, M.: Group recommendations: survey and perspectives. Comput. Inform. **33**(2), 1–31 (2014)
5. Kosala, R., Blockeel, H.: Web mining research: a survey. ACM SIGKDD Explor. Newsl. **2**(1), 1–15 (2000)
6. Kukar-Kinney, M., Close, A.G.: The determinants of consumers' online shopping cart abandonment. J. Acad. Mark. Sci. **38**(2), 240–250 (2010)
7. Mills, C., Bosch, N., Graesser, A., D'Mello, S.: To quit or not to quit: predicting future behavioral disengagement from reading patterns. In: Trausan-Matu, S., Boyer, K.E., Crosby, M., Panourgia, K. (eds.) ITS 2014. LNCS, vol. 8474, pp. 19–28. Springer, Heidelberg (2014)
8. Wang, W., Zhao, D., Luo, H., Wang, X.: Mining user interests in web logs of an online news service based on memory model. In: IEEE 8th International Conference on Networking, Architecture and Storage, pp. 151–155 (2013)
9. Wojewnik, P., Kaminski, B., Zawisza, M., Antosiewicz, M.: Social-network influence on telecommunication customer attrition. In: O'Shea, J., Nguyen, N.T., Crockett, K., Howlett, R.J., Jain, L.C. (eds.) KES-AMSTA 2011. LNCS, vol. 6682, pp. 64–73. Springer, Heidelberg (2011)
10. Tan, M., Shao, P.: Prediction of student dropout in e-Learning program through the use of machine learning method. Int. J. Emerg. Tech. Learn. **10**(1), 11–17 (2015)
11. Zhou, B., Zhang, B., Liu, Y., Xing, K.: User model evolution algorithm: forgetting and reenergizing user preference. In: International Conference on IoT and 4th International Conference on Cyber, Physical and Social Computing, pp. 444–447 (2011)

Framework for Managing Distinct Versions of Data in Relational Databases

Makarov Stanislav[(⊠)]

Don State Technical University, Rostov-on-Don, Russian Federation
nipheris@live.ru

Abstract. Despite the growing number of different database and storage technologies, relational databases still support significant percent of internet and intranet applications due to their effectiveness in processing complex queries and capabilities to provide data integrity in concurrent data access. Most of modern relational DBMS develop their concurrent access models to meet the requirements of transaction consistency, but at the same time they lack tools that provide usability in complex scenarios of concurrent data processing, that require storing and retrieving multiple versions of the same tuple. This paper describes the high-level framework to manage multiple co-existing versions of data tuples, and to support complex workflows of concurrent data modifications, including "branched" versioning. The different aspects of design and implementation of such a system are discussed.

Keywords: Relational databases · Data versioning · Data history · Conflict resolution · Rollback

1 Introduction

A lot of modern applications and information systems utilize relational databases for data storing and retrieving. Relational model became popular in consequence of the ability to model facts from the real world using the strong mathematical foundation. Other database models, such as document-oriented or graph databases became more popular recently, but RDBMS is still the highest-demand primary storage for the majority of today's financial, corporate or consumer applications that generate OLTP workloads.

In many information systems, the significant portion of data is created, edited and maintained by system operators - users, who are responsible to support consistency and completeness of some core data in the system (sometimes called *master data*). The examples are online store catalogues, digital libraries and scientific databases. While the rest of database can contain data, produced by automatic operations (such as order processing in online store or book reservation in library), that operations use human-authored directories and lists to reference to.

Due to error-prone and concurrent manner of manual data entering, there is a necessity for some tools coordinating the simultaneous work with data.

© Springer International Publishing Switzerland 2016
M. Ivanović et al. (Eds.): ADBIS 2016, CCIS 637, pp. 229–234, 2016.
DOI: 10.1007/978-3-319-44066-8_24

Such tools are well known and broadly used in document storages and software configuration management systems. It is *version control systems*, providing the ability to store and concurrently maintain distinct versions of some document or file. Version control tools like Git or Subversion play a critical role in software development process, but are tailored to work with textual and, sometimes, binary data. All of the supported operations, including merging and requesting of change history, are arranged to treat the file as a sequence of lines, or, as a sequence of bytes.

Considering the normalized nature of relational data, it is quite reasonable to handle the data using a tool, that preserves the benefits of normalization and allows fine-grained operations to be performed over normalized data. In this paper we observe some existing approaches and solutions to the problem of storing distinct versions of relational data, and outline the possible ways of improving this approaches.

2   Related Work

There is a significant effort to the problem of storing multiple versions of same object or entity in a database. Several papers addresses this problem in context of CAD/CAM systems. Klahold et al. [7] combined different known approaches (including previously existing version model for CAM databases) into general model for version management using the concept of *version environments*. Dittrich et al. [5] introduced the concepts of *object version* and *design object* (which groups multiple versions of one object together) for CAD databases. Biliris [3] enhanced this approach by definition of time as orthogonal dimension to versioning.

Other solutions come from area of object-oriented databases. Sciore [9] proposes a *user-level versioning* for database objects, where user-level versions are the object variants that are created for the needs of application and subject area (with contrast to system-level versions, that are created by DBMS to support concurrency and transaction processing).

More recent papers describe the versioning models for relational databases. First of all, this is large research topic related to temporal data support in databases. The recent update to SQL standard - SQL:2011 introduces the notion of *valid* and *transaction* time [8] with the later supposed to model the linear history of data modifications inside RDBMS. Solutions using transaction time work well in many practical scenarious, but the linear history, associated with timeline, is not powerful enough to model complex cases of multiuser data editing. The different approach, that enables branched versioning is proposed by Chatterjee et al. [4] and is implemented in Oracle Workspace Manager. Similar approach can be found in the implementation of ArcSDE subsystem of ESRI ArcGIS [6]. This subsystem is tailored to support long transactions and multiple co-existing versions of database to allow many people to perform editing tasks simultaneuosly in a geodatabase.

The common problem with solutions observed is the absense of unified vendor-independent and software developer friendly interface to work with versioned relational data. The another one is the scalability problem induced by using the relational storage both for vesioned data and version-related metadata. The highly-connected version metadata with lots of parent-child links doesn't fit well to relational model and requires execution a lot of select or join queries to get the version tree structure. The project described in this paper is intended to deal with this problems.

3   Multiversion Database Model

In this section we will try to build formal notion of multiversioned relational data. The basis we will use is the versioning model defined by Chatterjee et al. [4] and the similar model used by Git version control system [1]. Both of this models introduce directed acyclic graph of versions (or revisions).

Consider the database transaction $t_i \in T$, that changes the database state from s_{i-1} to s_i (let the state s of a relational database to be a mapping of each relational variable r_i to some relation R_i^s, i.e. $r_i \xrightarrow{s} R_i^s$). Notice that every such transation t_i has the only one database state to be based on and the only one derived database state. Such model illustrates the data processing flow in most modern relational DBMS. Even if several transaction are processed simultaneously, database system uses some of the concurrency control mechanisms to keep database state change sequential.

Now let's assume that every new transaction can use *any* of the database states that was effective earlier, and that some transactions can be based on *several* database states instead of only one. Consequently, our chain of states evolves to the directed acyclic graph of states $G = (S, P)$, where S denotes the set of states (every state is a node), and P is a set of arcs (s_i, s_j) where s_j is a some state, created by transaction t from state s_i. Notice that if the transaction uses more than one base state to proceed, each base state is connected to resulting state by the separate arc. In the example on Fig. 1, transaction t_5 is based on states s_3 and s_4 and produces the state s_5, so the state graph contains arcs (s_3, s_5) and (s_4, s_5) respectively.

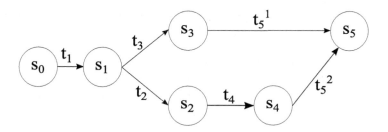

Fig. 1. Database state graph

The possibility to execute new database transaction over any database state existing in the state graph allows different database users to proceed with their transactions without the necessity to serialize them to the single linear sequence. Each user is able to create the separate transaction chain and to incorporate his changes and edits in independent manner. However, in most cases we should have an ability to integrate the work of different users. Therefore we need a mechanism, called *merge* transaction, that uses several base states instead of only one. In case of data conflict between different versions of tuple in the parent states, merge transaction allows to specify explicitly, what conflicting value of the attribute from parent states should present in the resulting database state.

4   Physical Storage Problems

Current implementations offer various approaches to store database state graph, described in previous section. Performance of operations and searches over this graph, including traversing through parents or finding all paths to specified node, is critical to build the responsive and scalable versioning system. All of the existing implementations, including Oracle Workspace Manager [4] and ArcGIS versioning [6] use the relational storage for the metadata that is the same as one for the versioned data tuples. Oracle Workspace Manager adds several auxiliary columns to the table when it is marked as versioned, and stores the workspace hierarchy in separate service table [2]. ArcGIS versioning subsystem makes two "delta" tables and uses them instead of the original one when performing queries. Also, there is an additional table to store the version tree and another table to cache the whole history, or "lineage" of the specific database state [6].

The problem is that retrieving of state hierarchy from the relational storage requires a series of join operations between the tuples representing the source and derived database states. The average quantity of the required joins grows along with the depth of the state graph. We propose to use different storage schemes to meet the requirements of versioned database - to store the history of tuple versions untouched and to have the limited cost of tuple lineage requests. Graph databases, that offer native graph storage, can provide the necessary services. Such databases use so-called "index-free adjacency", that means an ability to traverse between stored entities without involving index structures, leading to $O(1)$ complexity of single transition.

For example, consider the possible metadata graph on Fig. 2, that also includes a node for every primary key of the particular relation. Every link from the state node s_i to the key node k_j denotes the change done to the tuple with key k_j by the state s_i. If new tuple version is introduced by the database state (user INSERTs or UPDATEs the tuple), the link has a label "*STARTS*". If the current tuple version is considered to be retired by the state (user DELETEs the tuple), the link has a label "*ENDS*".

This structure provides some interesting possibilities to extract the information about each database state. For example, the problem of searching the current version of particular tuple with key k_j for the given database state s_i

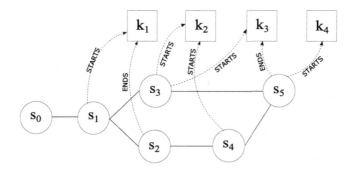

Fig. 2. Example metadata graph for the versioned database

can be traslated to the problem of searching the shortest path from the node s_i to the node k_j, and the conflict detection problem can be expressed in terms of finding all paths to the key k_j from the state s_i with different state nodes in the end.

5   Integration Problems

The modern information systems are complex things, and the database is the only one of the components. There are usually several applications that operate with data in the database and implement business logic. Many of the programming languages, that are widely used in development of enterprise applications are object-oriented. To keep the flexibility of relational data model, and the expressiveness of object-oriented language, developers use object-relational mapping tools to map the tuples of relations on entities in object-oriented model. In fact, the "virtual" object database is created and maintained by the ORM tool with the relational database as backend [10]. Such an approach has proven its efficiency in terms of developer's productivity and resistance to data model changes, and was adopted by many companies and projects.

Taking in account the idea of versioned database we mentioned before, we should consider providing some methods and tools to deal with multiple concurrent versions of data tuples, and to manage and resolve emerging data conflicts. Tools like Subversion or Git assume that user is familiar with the content of merging files, and is able to process them "as-is", without introducing the additional abstractions. In fact, file-oriented version control systems treat a specific file as an atomic block of versioned data, and make the user fully responsible for the *semantics* of merge operation.

In the discussed model of versioned relational database the atomic unit of data is the versioned tuple of a relation. In well-modelled and highly-normalized database, the data about specific domain entity can be spread across the tuples of several relations. Tuples of relations in the database often become too "low-level" in terms of representation of domain entities (because of using foreign and surrogate keys), and are not adequate to be directly processed by the system operator.

The idea we propose and develop is to provide the versioning of entities at object level and to perform merges and conflict resolution between object graphs. We are going to improve some existing ORM tools to allow them create and maintain a *versioned* virtual object database instead of a regular one, and to support all of the required mappings between the object view of data and the underlying relational representation.

6 Other Challenges

Some specific problems are also the subjects of ongoing and future research. The most significant are:

- estimate of the performance of version metadata storage based on graph database and the complexity of algorithms running over state graph;
- introducing the concept of *shortcuts*: the redundant specially crafted links in the state graph that are intended to speedup path searches at the cost of storage space;
- building and benchmarking a prototype of middleware system that implements multiversioned model interface;
- tweaking the underlying graph database to fit the structure of a state graph (high depth, small breadth, append-only lifecycle);

References

1. Git documentation. http://git-scm.com/doc
2. Agarwal, S., Chatterjee, R.: Versioned database system with multi-parent versions (2010)
3. Biliris, A.: Database support for evolving design objects. In: Thomas, D.E. (ed.) the 1989 26th ACM/IEEE Conference, pp. 258–263 (1989)
4. Chatterjee, R., Arun, G., Agarwal, S., Speckhard, B., Vasudevan, R.: Using data versioning in database application development. In: Proceedings of the 26th International Conference on Software Engineering, pp. 315–325 (2004)
5. Dittrich, K.R., Lorie, R.A.: Version support for engineering database systems. IEEE Trans. Softw. Eng. **14**(4), 429–437 (1988)
6. ESRI: Versioning. an esri technical paper (2004). http://downloads.esri.com/support/whitepapers/ao_/Versioning_2.pdf
7. Klahold, P., Schlageter, G., Wilkes, W.: A general model for version management in databases. In: Chu, W.W., Gardarin, G., Ohsuga, S. (eds.) Very large data bases, pp. 319–327. Distributed by Morgan Kaufmann Publishers, Los Altos, CA, USA (1986). http://www.vldb.org/conf/1986/P319.PDF
8. Kulkarni, K., Michels, J.E.: Temporal features in sql: 2011. ACM SIGMOD Record **41**(3), 34 (2012)
9. Sciore, E.: Versioning and configuration management in an object-oriented data model. VLDB J. **3**(1), 77–106 (1994)
10. Xia, C., Yu, G., Tang, M.: Efficient implement of orm (object/relational mapping) use in j2ee framework: Hibernate. In: 2009 International Conference on Computational Intelligence and Software Engineering, pp. 1–3 (2009)

Author Index

Printed in the United States
By Bookmasters